CHANGING SEASON

CHANGING SEASON

· ·

A FATHER, A DAUGHTER, A FAMILY FARM

DAVID MAS MASUMOTO WITH NIKIKO MASUMOTO

Heyday, Berkeley, California

Library of Congress Cataloging-in-Publication Data

Names: Masumoto, David Mas, author. | Masumoto, Nikiko, author.
Title: Changing season : a father, a daughter, a family farm / David Mas
    Masumoto and Nikiko Masumoto.
Description: Berkeley, California : Heyday, [2016]
Identifiers: LCCN 2016010026 | ISBN 9781597143660 (pbk. : alk. paper)
Subjects: LCSH: Masumoto, David Mas. | Masumoto, Nikiko. |
    Farmers--California--Del Rey--Biography. | Farm life--California--Del Rey.
    | Family farms--California--Del Rey.
Classification: LCC SB63.M36 M37 2016 | DDC 630.9794/82--dc23
LC record available at http://lccn.loc.gov/2016010026

Front Cover Photos: Korio Davis Masumoto
Back Cover Photo: Staci Valentine
Book Design: Ashley Ingram
Printed in East Peoria, IL by Versa Press, Inc.

Orders, inquiries, and correspondence should be addressed to:
Heyday
P.O. Box 9145, Berkeley, CA 94709
(510) 549-3564, Fax (510) 549-1889
www.heydaybooks.com

10 9 8 7 6 5 4 3 2 1

To my mom,
who worries every season.
You will always be part of this farm.

—D.M.M.

To my jiichan;
I miss you every day.

—N.M.

CONTENTS

Part One:
THE STRUGGLE OF CHANGE: Past Lessons

Part Two:
HOW WE LEARN: Present Tools

Part Three:
THE FUTURE OF FOOD: What's Next

PREFACE

David Mas Masumoto:

In 2013, the Center for Asian American Media asked if my family would be interested in participating in a documentary film. Naturally, my first response was, "About us? Really? A film about a farmer? Who would want to see that?"

We weren't that special. We didn't have a dramatic story. An evil bank was not foreclosing on our farm, and we were not on the verge of financial collapse. The drought had not wiped us out; we were not displaced Great Depression–era Migrant Mothers. No one in the family had a tragic past or a heart-warming tale of survival that could be picturesquely set on an organic farm with morning dew kissing the leaves. We are not *Duck Dynasty* with Asian faces. None of us has committed a major crime. No Ponzi peach schemes here. Marcy, my wife, was not leaving me; there was no wild love triangle, no sordid sagas of murder or blackmail. This was not farming in Fargo.

A film about us would lack the dramatic "pop" and flashy hook audiences have come to expect. All we had were our stories. I concluded that the proposed film would not be about us but rather about the farm, about real dirt. The stars would be the peaches—the alluringly fat and ripe peaches that invite foodies to bite into their sweet flesh and feel the juices slowly slide down their cheeks to dangle on their chins. This could work. Sex sells, and our fruits would be movie stars: food porn from the Masumoto family farm.

I was wrong. The film ended up being about us after all. We

exposed our lives, our history, our family, our struggles and challenges and smiles and laughter. It was about the everyday and the authentic. The film was anchored in story, a natural extension of what I have been doing for much of my life: writing. A major part of the unscripted narrative was the return of my daughter, Nikiko, to take over the farm, her voice and identity as a queer woman becoming part of the fabric of our family. They titled the film *Changing Season on the Masumoto Family Farm.*

Think of this book as an extension of and companion to the film. Each complements the other, but they both also stand alone. Each is born from a common desire to explore real lives and real stories.

The film took two years to make, including thirteen visits to the farm and other venues, each lasting multiple days over the course of all four seasons. This book was written over a period of six years, much of the material drawn from my monthly newspaper columns for the *Fresno* and *Sacramento Bees.*

After most of the filming was complete, I had a brief talk with the film's director, Jim Choi, and the editor, Chihiro Wimbush. They mentioned the term "cinéma vérité," but refused to elaborate, fearing too much explanation might ruin the authenticity and spontaneity. Only later I learned this was a documentary filmmaking method that allowed the story to unfold without a narrator—the camera recording real events and actual persons without directorial control. It's very similar to how I write, and how I approach life. In fact, it reminds me of our farm business plan: we never had one. We fell into organic farming in the 1980s, well before organic markets were established, because it felt like the right thing to do. As the organic community matured, we were in the right place at the right time. Luck. Good fortune. Good karma. No script.

I have always written creative nonfiction, and most often in the first-person voice. I use the word "I" a lot. This is how I see the world. I can't make things up—a tough constraint when reality doesn't make sense. I'm stuck with the truth. I can't make myself taller or wiser. I can't change my past, nor do I want to alter the history of people around me; we're bound by the ties that bind us.

I've also learned that the process of making a documentary is similar to how I approach writing. We explore. We probe. We understand. We're stimulated by the real world. We are revealed.

Changing Season the book is divided into three sections: the past, the present, and the future. Nikiko's voice figures in each section, her "field notes" adding another perspective, a fresh point of view. We are lost and confused; we question, challenge, grow with self-doubt; we sense a maturity with the passing of years. We share the same journey, a generation apart, marked by different realities. And we're still trying to figure it all out.

I'm not sure when the documentary began to take shape in the filmmakers' minds, but this book has been within me for a long time. The words you read here gradually evolved from a collection of stories into a narrative exploring traditions and change. And, hopefully, legacy.

If we're lucky, the reader and the film viewer will judge us kindly. We hope they don't leave asking, "Why did they make that film/write that book?" Instead, they will feel that stories can indeed fill a book and complete a film.

Nikiko Masumoto:

I realize that we Masumotos have a tendency to reach toward the profound. Maybe it's part of living in the legacy of family. Maybe it's part of the essence of who we are. Whatever the reason, I hope you find something earnest in our reflections.

My life has been so full of powerful moments experienced while living and working on my family's farm. Sometimes the abundance of wisdom is so large that laughter and tears seem like best friends. I have walked in our fields in silence so full I can feel my inhale reach my toes. I have stood on a tractor with my arms extended in victory, whooping and hollering to myself after accomplishing some small feat, like making a difficult maneuver with a large piece of equipment, or avoiding getting stuck in the mud. These experiences have changed and challenged everything about me. But why should I share these stories?

I feel compelled to share. On a personal level, the practice of crafting stories and transforming experiences into narrative is a fundamental part of how I understand my own life. More importantly, I feel an honorable responsibility to share because what we do is linked to so many lives beyond our own. While our stories are grounded in the specifics of our family and our farm, I hope they resonate wider, building bridges between our differences. I hope they inspire introspection and connection. I must believe that all of the gifts I have received from this life are not only worthy of sharing but must be shared. The food we grow already connects us as farmers and eaters; our consciousness and our humanity should follow.

When the documentary film project *Changing Season* started, I felt a new sense of anxiety. We were opening ourselves to a crew of filmmakers to document and retell something about our lives. What would they see? What would they say? And what would I see of myself? Filming was an intimate process that included beautiful moments of trust as well as negotiations of power and confusions of communication. Eventually, however, the anxiety gave way to the comfort of deep relationships. This film is a representation of our lives captured, stitched, and composed by people who started as strangers and ended feeling like family. And as we all know, family doesn't just mean the good parts, it means the whole complicated package of relating to, listening to, struggling with, understanding, confusing, growing, and accepting each other. As you read these reflections about our family farm, I hope you find something that makes you feel like family too.

PART ONE

THE STRUGGLE OF CHANGE: PAST LESSONS

COMING HOME: THE PATHS WE MAKE

My grandfathers left Japan and their families' small rice farms in Kumamoto (my father's side) and Hiroshima (my mother's side) because they had no future. They were cursed as second sons, destined to live lives of poverty. They were trapped in an old hierarchical network that meant they were not in line to inherit their family plots. To make matters worse, in the late 1800s and early 1900s, agriculture in Japan had stagnated. Taxes were high and prices depressed. There were no startups in rural Japan and little opportunity or reward for entrepreneurial spirit. Immigrants came to California with few other options. They ventured to this new land, a world of production agriculture that needed strong backs for cheap labor. But they were different: Asian aliens who did not believe in God, spoke a foreign language, and were not white. They wore the wrong faces for America.

Many of these Issei (first-generation) men convinced potential wives back home in Japan to journey to the United States as picture brides—their marriages arranged on little more than an exchange of photographs. Some say these women were often tricked, photos of their future husbands switched for those of younger men in front of big farmhouses, creating an illusion for the unsuspecting partners. But my grandmothers, uneducated and unsophisticated, were themselves escaping bleak lives as farm women from poor families in rural villages, and perhaps the deceptions were mutual. In some ways, it was an eighteenth-century version of online dating, except you had to cross an ocean for the first meeting, with no opportunity to turn back.

My family arrived in America with low expectations, and they accepted the hard physical work. They came with no intention of returning to Japan. Immigrants had no issues with succession and the transfer of knowledge. Everything was new and original and foreign. They grew grapes and tree fruit instead of rice. They remained strangers in their adopted land.

From 1942 to 1946 our family, along with 110,000 other Americans of Japanese descent, were imprisoned in relocation camps scattered in desolate areas of the United States—all because they looked like the enemy Japanese who had bombed Pearl Harbor. Dad spent three years living behind barbed wire—angry, bored, and hungry to live. When the United States entered World War II, he had just graduated high school in California and the family was preparing to purchase land. Instead, they were uprooted and exiled to the internment camp at Gila River, Arizona, without having committed any crime, displaced because they were supposedly not American. To add insult to injury, my father, after years of incarceration, was drafted into the US Army.

Dad shared with me his stories of coming home after the war. When he returned to the Fresno area in 1946 after his stint in the army, the family was broken. His oldest brother (the "number one son") had been killed in France fighting for the country that had locked up his family. Another brother also had been drafted and was stationed in Europe. A sister and brother had left for the Midwest, leaving two old non-English-speaking parents and a young teenage sister behind, still incarcerated. Our family was one of the last to leave the internment camps, because they had no place to go. They returned to Selma, California, and for weeks they slept on the floor of a closed grocery story once operated by another Japanese American family who also had recently returned from the Arizona desert.

On my father's first morning back after having been discharged from the military, my grandmother shook him awake and said he had to find some farm work for all of them. They were penniless, hungry, and desperate. The Torri family needed to open their store. The Masumotos would be without shelter. Welcome home.

For two years, Dad scrambled. He found a barn the family could live in while he worked. Later they moved into a shack on rented land. Dad worked weekdays on other farms, tending his own fields after hours and on weekends. He understood that land-owners were rewarded, not farmworkers. He took a huge gamble and bought a piece of dirt that was full of hardpan rocks and roll-ing terrain but was cheap.

He broke the news to my grandfather, who, then in his sixties, was old, worn out, and emotionally crushed. My grandfather was happy that his son had realized the family dream of owning a farm, and he quickly packed what little he had and shuffled to the vehicle. My father readied the car to take his parents from the shack they'd been living in.

But my grandmother was furious. "Why you gamble?" she argued. "Crazy. They take things away from you in America," she blurted. "You buy a farm like it was a sack of rice. You make mis-take. Big mistake."

They argued, and Dad gave her an ultimatum: he would wait in the car until sunset and then go to the new place, with or without her. He and my grandfather sat waiting. Finally, with a brilliant setting sun in the background, my grandmother trudged out carrying a small black suitcase with the number 40551—the Masumoto relocation identification number—stenciled on the side. Together, in silence, they drove to the new Masumoto family farm.

When my time came, I also left and came back. Like most American farm kids in the sixties and seventies, I fled the dirt and sweat and hard work, encouraged by parents who wanted something better for us. "Get out of the fields," they advised. "Find something else—a safe profession." And most of us listened.

I ran from the farm and family as far as I could, figuring UC Berkeley and the San Francisco Bay Area's radical ideas would take me away to another world. For two years, from 1972 to 1974, I lost myself in a swirling sphere of politics, my search for an Asian American identity, and the fast pace of the big city. I returned home for the summer between my freshman and sophomore years and told my father this would be my last harvest. He listened and talked of pulling out our peaches and nectarines; raisin grapes, he figured, didn't require as much work and management. (Years later, I regretted that conversation. It was a stupid, immature thing to have said "I'm never coming back to this farm," and it probably broke my father's heart. But I was playing my electric guitar, listening to Led Zeppelin, believing I was revolutionary, seeking my freedom.)

Yet even while at Berkeley, I still hungered for more, so I journeyed to Japan to study for two years. I wanted to put thousands of miles between my two lives, and I immersed myself in another land, another language, another culture, far away from a farm, from California, from America. Tokyo churned with a new and different life with its millions of residents, its compact city streets, and its sleek trains. The bright lights of the nights and the constant rush of energy exploded before me. A new world beckoned, inviting my imagination to soar. I ran through the crowded streets and listened to the voices of this different land and her people.

After a few months, I took my first trip out of the city. To save money, I took a series of local trains instead of the express. We stopped at all the smaller stations and slipped into the exposed spaces of the countryside. I was stunned by the openness of rural Japan, with which I felt a strong affinity. I decided then to plan a trip "back home"—to the village in southern Japan from which my grandparents had emigrated.

A few months later, I found myself venturing back in time to a rustic farm village outside of Kumamoto. Awkwardly, I met my grandmother's brother at the train station; he did not understand my polished "Tokyo Japanese," and I could not comprehend his Kumamoto-ben "southern" Japanese dialect. We drove in silence along country roads and arrived eventually at the small village of Takamura. My grandparents had left sixty years earlier, and no Masumoto had since returned.

I was welcomed. I slept and ate in a centuries-old farmhouse. For weeks, I worked the ancient rice paddies of my grandmother's family. I breathed the country air, listened to the sounds of nature and the farm. I identified with the patterns of farm life and felt at home. The very life I sought to escape I had rediscovered. This was not the coming-of-age story I had planned.

Yet, as I shoveled weeds and helped plow the fields, I could not find the right farming rhythm. The pace of daily life there centered around rice, an unfamiliar crop that I had no sense of how to work. Though I had eaten rice daily in America, I was clueless about how to grow it.

Returning to Berkeley, I felt lost for most of my senior year of college. They call it "reverse culture shock" when you return from a foreign land and still feel alien at home. My angst cut deeper than that, though. While in Japan, I had foolishly hoped to maintain a relationship with a girlfriend in California. I was wrong and wronged. After a few months, she had stopped writing; promises of young love were shattered and my heart was broken. My life sounded like a bad country-western song: "She took my heart and stomped on it." I tried writing my way out of depression. I composed bad poems. I was pathetic.

But something else pulled at me. I felt a tug back to family and the farm. I felt an obligation to take care of my aging parents and grandmother. I wanted to repay something to them and hoped they would allow me to return, at least for a while. I needed to sort things out while I peered inward and did some soul searching.

And I dreamed of writing. I put aside my bad poetry—crappy, bad writing due to crappy, bad thinking. Creative nonfiction stories now attracted me; I was gradually drawn into the real. I secretly wished to become a writer of community. A keeper of oral traditions and histories. A writer who lives, works, grows, and loves in the same place for decades. An insider carrying stories and the burden and responsibility to share them with the outside.

At first, I thought the two might make a good match: farming had an off-season that would allow me to write. I assumed that beginning writers would not earn much money, so farming might fund my dream. I quickly learned that neither farming nor writing made much money, but in my optimism I rationalized that this made them even more compatible. I created my own personal nonprofit movement.

My homecoming was complete when I started farming organically and made a family. I married a farm girl. Marcy was not a Sansei (a third-generation Japanese American) like myself but a woman from Wisconsin Catholic/Lutheran stock who had grown up on a goat dairy in the high deserts of Southern California. Even though she was actually born in California, she probably grew up more Wisconsin than those actually living in Wisconsin—a common trait of the children of transplants with entrenched identities. Her family named the California street they lived on "Wisconsin Street."

Organic farming guided our trajectory. I wanted to take care of the planet, and I hoped to build a new network of farmers that would help change the established, centralized systems of food production. The birth of our two children, Nikiko and Korio, completed our circle. They were raised in a mixed-race household with a blend of Buddhism and Christianity, and they learned to farm with new traditions with my parents nearby (though having at least separate driveways was important to Marcy).

Nikiko is our oldest child and was born with a love of the land. You could say I stayed home to raise her through her infancy, but that's not quite true; I took her out into the fields to work with me. She'd take naps in the cab of an old pickup or play in the dirt while I shoveled nearby. She was vocal and studied the world around her with intensity. Growing up in a natural world full of more mystery than simple solutions made her crave explanations. Nikiko never stopped asking questions.

Every generation came to this farm with its own set of circumstances. Each carried baggage of anxiety, uncertainty, and naiveté. Each began anew and came to understand and appreciate change. None of us were pioneers—farmers are not typically the first in anything. We are settlers who want to stay put, to marry a place and plant a part of ourselves in the earth. Fate willing, we will pass down what we have learned and die on our own land.

NIKIKO'S FIELD NOTE: CHOOSING A LIFE

I came to this farm three times. The first was without my consent (although none of us chooses where or into what environment we are born), but the next time, I was lured by a radical dream, and the third I returned with a longer view of life and tools. This last time, I'm pretty sure it'll stick.

I was born into a lineage of family farmers. I am the first child of two pioneers in organic farming: my dad, David Mas Masumoto, and my mom, Dr. Marcy Masumoto. Both my parents grew up in agriculture, their parents also living tied to the earth. Our home was a white wooden house floating in a sea of green: hundred-year-old grapevines and orchards planted in my father's lifetime.

Now, I have come to appreciate the leap of faith they took at the time of my birth. They were newlyweds, cultivating a mixed-race love together on a farm they were transitioning to organic. And this was before organic was hip or profitable. There was no guarantee the experiment would work, and in fact, the year I was born, 1985, the farm lost money. I didn't choose this childhood, and while I enjoyed the closeness of working with my family and the incredibly juicy peaches we raised, I did not want to be a farmer.

In school, I remember other students making fun of a class-mate, John, who was also a rural kid. They mercilessly called him "Farmer John" and mistreated him, assuming he was dumb just because he lived in the country. (Of course, the irony was that most of us lived in rural settings.) The social equation was clear: farming was not a smart or cool choice. By the time I reached high school, I wanted to leave the farm and the Central Valley of California for other reasons too. I felt confined, and, full of willful teenage ambition, I thought myself ready to burst into adult life. In my imagination, success always existed elsewhere, not here, not in this place of my birth—a rural place, a farming community, a conservative Central Valley, an exit off Highway 99, just a place that people pass through. How small home felt.

But then something happened in my second year at UC Berkeley. I took an environmental studies class in which a guest lecturer, Dr. Susan Kegley, talked one day about pesticides. She examined their impact on a global scale in terms of their toll on human health and the environment. She showed slides and referenced studies documenting the risks and ramifications of living in a world where pesticides ruled supreme. Suddenly, the place I had grown up in looked different. The organic farm I ran away from and the work my parents had been doing all along—the work to bring back life to our fields—seemed courageous, political, and even radical. In a global context, I saw my parents as pioneers of a better way of living.

At the time, I was majoring in gender and women's studies. My soul had already been set ablaze by my enhanced ability to consider a wider range of perspectives, to recognize and articulate how power functioned, and to dream of new ways of thinking, acting, and being free, equal, and empowered. With my passion for justice, I realized that organic farming could quite possibly be one of the ways my family was radically transforming the world in one small eighty-acre corner of the planet.

I revisited some of my dad's early writings and saw that he was enacting critical race theory in the fields by first being conscious of and then articulating how race, ethnicity, and history frame our family farm. I saw our small bit of land as a place of revolutionary departure from systems of dominance. Mom and Dad, did you realize then the power of what you were doing?

By the end of my time at UC Berkeley, I had decided to come back to the farm, and I returned with a vengeance.

When I first arrived, I thought I was ready. I had big dreams of starting a momentous movement for freedom and equality in my home communities. I came with a heart ablaze, but I soon found out I was burning too quickly. I met the valley with protest, in hopes that it would respond with change. But I was lacking sufficient tools, and I was brazenly impatient. The fundamental roots of my grit were underdeveloped and needed time and experience to mature. In order to learn to farm and to build a home and community, I needed patience coupled with creative thinking and action. So I left the farm a second time to explore the juncture of performance and memory, tending to the artistic yearnings of my heart but always thinking I would come back eventually. Just not quite yet.

After earning an MA, I came home a third time, now with a larger understanding of myself in the world and with new strategies of creation in addition to my old passion for protest. In the meantime, the farm itself had also changed; my jiichan/grandfather had passed away during the first year of my master's program. Death changes everyone's role on the farm. The younger generation moves up in line. Perspectives shift.

I came home with tools I had learned in graduate school—skills to manage time, a deeper emotional intelligence (although this is something that will grow forever), and a thirst for work that engaged my body (we so easily forget in academic work to honor and attend to the body). And, most importantly, I came home with a more patient view of my work unfolding across a lifetime. I had learned about new models of creating change that depended on my sustainability too. I knew grit and patience worked together: grit being the ability to work devotedly for extended periods without much reward, and patience being the awareness that some things take time, and how much is not always clear.

I came home not knowing everything, but wanting to learn. And so began my third beginning, a returning and revolving. I thought, at last, I wanted to be a farmer. Little did I understand what that really meant.

WHEN THINGS BREAK

When I was very young, I snapped the handle of an old shovel. This wasn't any old handle but one that had been worn smooth by years of use in my family. The grain was shiny and cool to the touch. My grandmother and father had worked with this tool, making countless strokes that wore the blade down from a point into the gentle concave curve of a half-moon. They never left it outside at night but instead brought it into the shed and leaned it against the wall in its special place, as if it needed to rest and rejuvenate before serving us well one more day.

As a twelve-year-old, I had been rushing to dig a hole around one of the cement irrigation valves in the orchard so Dad could patch a leak in the concrete. The dirt was soft, but tree roots had wrapped themselves around the concrete cylinder. The hole was supposed to be at least three feet deep, but I never got that far. In my haste, the metal shovel blade hooked a thick root and I pushed against it with all my weight. I felt the handle bend and then heard a subtle yet audible crack escape from the wood.

I ignored the warning and pushed harder. The wood snapped and gave way as it fractured at the base where it connected with the metal face. The blade hung lifeless, dangling sideways. I limped home to my father. He should have been mad with his impatient son, a worker using brute force rather than common sense and discretion. Instead, he said nothing. And he did not toss the old tool aside and buy a new one. He studied the broken handle, nodded, and pulled out the surviving handle with the splintered end. Then he began to shorten it. We are not very tall people, so a compact shovel could serve us well. He would fix my mistake with a simple remedy. No yelling. No sign of disgust.

I wanted my father to be angry, to show emotion. A child grows up seeking recognition, forgiveness, and acceptance, even when he breaks things. Did my father even care? Should I care? My guilt morphed into shame.

Another time, Dad taught me how to drive a tractor while pulling a disk. I was proud to break through the topsoil crust, slice the earth, rid the fields of weeds. But at the end of one row, I turned too wide and didn't center the equipment for the next pass. I tapped a cement irrigation valve with the outside disk blade, and I didn't stop. I ignored the accident. I hoped it would go away. Maybe it would even fix itself. A classic hit and run.

I conveniently forgot about the incident until Dad started to irrigate the vines. Instantly water gushed from the severed valve, flooding the immediate area, stealing water from other rows. I then confessed. Well, sort of. I stammered out loud that something might have hit the valve, that it might have already been broken.

Water is the lifeline to our trees and vines. Dad dug out the crack and worked his cement art. It took days for the patch to set, harden, and dry, and we had to delay irrigation.

Dad said little, offering no sage words about being careful, not even a brief lecture about taking care of things before they're needed. No sermon about avoiding a crisis. I was forced to witness in silence as the thirsty leaves drooped for a few days, longing for a drink. I was confused by my father's silence; was this my punishment for lying?

Over the years, I've knocked over plants, disked too close to trees and sliced some roots, and hooked many vines with a French plow. Each accident left physical evidence, a place marker. My instinct was to replant the damaged vine or tree: fix and replace. And forget. Dad, however, would try to save the fallen creatures. He'd push them back up, fashioning wood braces to support the leaning trees until new roots could be established. Most grew back, albeit weaker. They survived as a constant reminder of... what? Was he teaching me a grand lesson? Was this something I might one day understand, but only after a lifetime of regret?

─────────────

My father was not a storyteller. When I came back to the farm after going away for college, I worried how I could learn from his lifetime working this land. How would he pass down what he knew if he never talked about it?

Dad was a quiet and reserved man. He spoke louder with his tools and work gloves than his voice. We'd labor side by side, little said, and often I'd hear only the tramping of our boots trudging through the fields or the buzz of an arc welder as he repaired farm equipment.

A silence filled our world. A silence that spoke of a history of working the land and the challenges of being poor and Japanese American. A silence that grew over the years to fill the void between generations.

I still think of this silence as I work the same lands with Nikiko, who will be taking over the family farm. But we are different. I talk a little more than my father and love to tell stories. Nikiko asks questions and doesn't hold back her passionate opinions and emotions. Yet we too share the wisdom of silence as we work the land.

How do I transfer knowledge? How can I share my experience and lessons with my children? Should I contribute my thoughts on all occasions, or are many things better left unspoken? Farming, which may appear unchanging to an outsider, is undergoing a huge upheaval. The business of growing what we eat, the rapid introduction of technology, and a shifting culture of food are transforming our land and the way we work. Every generation claims that the times are changing—children think and act differently than their parents, and the old must give way to the new. I remember becoming impatient with the ways of my father, believing his archaic methods needed to be replaced by something different, something better. To do something different is to tell a prior generation they are no longer doing it right.

Now that I have aged, I know I was partially right, and also wrong. The old and the new both have a place in our lives. Traditions sometimes survive because they work, and yes, information needs to be passed down, but at the same time, I must be open to new approaches and directions. Simultaneously, both parties will be impacted and altered, but without a transformation, there is no transfer of wisdom. No one sets out to become a fossil.

I believe in the power of experiential memories. When information is truly learned, it's not a matter of memorizing facts and formulas. When experience is mastered, something transformative happens: memories become stories.

In the end, just like my father, I don't always know which are the most important moments and lessons to pass on. Wisdom should be fluid. My life has been a collection of dots, experiences that only made sense and became significant afterward, when I could see them as part of a larger picture. Connecting these dots can only be done later in life, as we mature and come to better understand context and process. But it is confusing when we are in the moment; solutions are rarely revealed in the present. And in the future, I will continue to break things.

Both silence and stories echo throughout the fields of our family farm. We grow perennial crops here—not just peaches, nectarines, and raisins, but also family and history and more stories to tell.

SUKIYAKI

The summer of '63, I was a nine-year-old farm kid. My rural school had about 250 students, my class about 30. I spent summers working on the farm, no differently than all my friends. Outside our little world, change was churning, but here life was about peaches and nectarines and 100-degree days. I was unaware of the heat off the farm and in the cities. The world was becoming a new place.

Fundamental political transformations were sweeping across our nation. Civil rights battles overflowed into the streets, and confrontations and killings sparked racial unrest. A Baptist church was bombed in Alabama, killing four young girls. At a "March on Washington" rally, Martin Luther King delivered his "I have a dream" speech. The Viet Cong claimed their first victory over South Viet Nam and American forces. The Supreme Court ruled that institutionalized prayer in public schools was unconstitutional.

But for me, the farm kid, what I most remember about 1963 was that the top musical hit that summer was sung in a foreign language and blasted everywhere on our radios. The song wasn't in Spanish (Ritchie Valens's "La Bamba" was released in 1958 but never reached the top spot, although an Italian-language song called "Volare" did hit number one that year), and it wasn't in French (although later in '63, the Singing Nun and her

"Dominique" did shoot to the top of the charts). No, the musical hit of the summer of 1963 was sung in Japanese. "Sukiyaki" reigned at number one for three whole weeks.

I recall first hearing Japanese on the radio when my non-English-speaking grandmother listened to a local Sunday-morning show with news and music. A Buddhist minister sometimes shared a sermon or chant, and my tiny baachan/grandmother would put her hands together in gassho at the end of the talk.

But most of the time, our radio was tuned to KYNO, the local pop music station that played the top hits aimed at a teenage audience. We listened to it all day during the summer, in our farm's fruit packing shed and in the fields on little transistor radios tucked in our shirt pockets.

When I first heard "Sukiyaki," I was shocked to hear a Japanese song echoing in our shed. Japanese on a rock-'n'-roll station! Japanese that I could be proud of instead of embarrassed by. And it wasn't just a hit with us—an entire nation of young people seemed to like the beat, enjoy the rhythm, and accept the words. My school buddies mimicked the lyrics (albeit badly). I sang with a poor Japanese accent, faking most of the words. One of my favorite parts was the whistling—that lost art of making music with one's lips. I spent much of that summer whistling in the fields—both "Sukiyaki" and "Whistle While You Work," from the movie *Snow White*—testing the premise that whistling makes work easier and time go by faster.

An import from Japan had penetrated the American landscape and risen to number one. Even here on the farm, I felt something different in the air. Maybe change could happen in all of America, even out in the sticks.

For Japanese Americans, the success of "Sukiyaki" symbolized a type of acceptance. Only twenty years prior, my parents and grandparents had been interned in relocation camps following the bombing of Pearl Harbor. They had been deemed the enemy based on their appearance. For four years, tens of thousands of Japanese Americans—many American citizens by birth—were sent to prisons in desolate areas of the United States because the government considered them a security risk. Our family was forced to call the Arizona desert home during that time, and we lived there behind barbed wire in bleak wooden barracks shared with others.

I was born after this tragedy and heard very little about it from my family and community. They internalized the pain and suffering, the shame and disgrace of false imprisonment, and the abandonment by their own country. Yet after the war they returned to their American homelands and remained silent and stoic as they worked their way back.

Then came "Sukiyaki," a song in Japanese about lost love and looking to the future. The song had a timeless sweetness about it, a soft melody and upbeat rhythm with a message: to smile amid hardship, to make the most of your life. One repeated line, "I look up when I walk so the tears won't fall," seemed to capture the Japanese American ethos to look beyond the past, accept the sorrow, and move on. "Sukiyaki" was about lost love, about remembering the happy times even as you feel alone. The singer looks up and counts the stars with tearful eyes but knows that happiness lies beyond the clouds, above the sky. So look up and the tears won't fall.

Of course, few of us Sansei, and most of the rest of America, didn't understand the Japanese lyrics. The majority of listeners didn't know the title was changed to "Sukiyaki" for an American audience because the original, "Ue ō Muite Aruko ("I Look Up as I Walk"), was too hard for English speakers to pronounce. Sukiyaki is a type of Japanese stew that has nothing at all to do with the lyrics of the song.

People didn't realize that the songwriter had written "Sukiyaki" in response to the failure of Japanese protests against American military presence.

My uncle loved "Sukiyaki." He was fluent in Japanese and English, had served in the US Army in the Pacific as a translator during World War II, and was constantly trying to balance his love of both countries. I grin when I imagine him, a fifty-year-old Japanese American man, going into the local record store and proudly buying a 45 of the top summer hit.

At this time, Japan was still rebuilding after the war and was about to reenter the world stage. A spirit of starting anew swept that nation and spread across the ocean. Japanese Americans adopted a similar spirit of renewal, despite the loss and sadness of their personal histories: it was time for a fresh start. Good fortune was beyond the horizon. Look up and the tears won't fall. We all seemed to set our sights forward, not backward, searching for something, but not sure of what. When I grew older and had a family, would my children inherit this spirit of optimism, this determination to endure the past and search for something more?

"Sukiyaki" helped my family. It helped them begin to lose their sense of shame following their wartime relocation. My parents smiled when they heard the song on the shed radio; perhaps they were no longer humiliated, perhaps their wounds were healing. That summer, our peaches seemed a little sweeter. Prices weren't that bad, and it felt as if maybe the world was accepting us a little more.

Many still harbored ill feelings against the Japanese following the war, but they were aging. Sometimes hate, racism, and prejudice does die and go away. Could a pop song have the power to change cultural and social definitions? Or was it an indicator of shifting attitudes? A new, younger America was moving on, accepting a Japanese song and lyrics by making it the number-one song in America.

For a Japanese American farm boy in 1963, that song meant a new world was open to me.

NIKIKO'S FIELD NOTE: WEST SIDE STORY

When I was a little girl, I was fascinated by musical theater. When I got chicken pox in fourth grade the day before I was to see my first live musical production, I lied about my symptoms, went to school anyway, and hoped that I could still go to the theater. My brilliant plan was outplayed, however, by the conspicuous bumps and sores that emerged all over my body, and not only did I not get to go to the theater, I also infected my whole class. Whoops.

But my day did finally come. My mom took me to see *West Side Story*. I was so excited.

I craned my neck to see over the guardrail in the front row of the farthest balcony in the William Saroyan Theater in Fresno. I wanted to catch every kick, dip, and finger snap. Though we were sitting so far away that I couldn't see faces clearly, I didn't care. I imagined it was Natalie Wood and Rita Moreno dazzling on the stage.

I remember the theater lights jumping from skirt to skirt during the dance scenes as the epic mixed-race love story unfolded in the context of a 1950s urban landscape. I was mesmerized by the choreographed drama whirling in front of me; I was watching real bodies, real moving and breathing, and I felt my heart swell and tear with real emotions. It was the first live show I had ever seen, although it was not the first time I had encountered star-crossed lovers Tony and Maria. By nine years old, I already knew the story well. Except for the ending.

I watched the movie version of *West Side Story* countless times at home on Saturdays (the only day I was allowed to watch TV). I knew practically all the lyrics, I knew when to start snapping with the Jets and the Sharks, and I danced in my living room, proudly declaring a proto-feminist space for my own beauty along with the heroine. Yes, Maria, I feel pretty too!

The version I watched had been taped from a broadcast on television. In the days of VHS, we had economically creative means of participating in pop culture, and in this case it meant my dad had figured out how to record television shows onto VHS tapes. I remember the rush and the anxiety of watching my dad stand by the tape player, finger hovering, waiting to push the right button at precisely the right moment to start and stop a recording that would omit the commercials. The process was usually a team effort. Those of us watching would yell, "Now, now, now!" and he'd push the magical button. Many of my most treasured cinematic memories were captured in this manner.

There were, however, several risks to this method. First, you could hit "record" too early and a small excerpt of commercial would be spliced in between scenes, or you could start recording again too late and miss a small part of the next scene. While a little annoying, neither error dramatically affected our viewing, and even the mistakes have worked their way into the realm of my nostalgia. (I still have a couple seconds of an Isotoner glove commercial burned in my memory.)

There was yet another way to mess up taping from television, and this one was the worst: you could run out of tape. That's precisely what happened with our "borrowed" copy of *West Side*

Story. In our version, the story stopped right after the big rumble. Bernardo and Riff, the leaders of the gangs, were already dead and, out of revenge, the character Chino was searching for Tony to kill him. The last scene I saw was a close-up of Chino in the streets in the middle of the night, his face sweaty with rage and sadness, his hands gripping a gun.

When my mom took me to the live show, I was excited for so many reasons, but the most important was that I had waited all my life to see the ending! Throughout my childhood, my imagination had created many possible finales, and I couldn't wait to see which one was true, which one was real.

When the live play came to a close, I could barely see the stage. Massive tears had turned everything into a watercolor. I understood what had happened, and I was in shock. I don't recall if I even clapped. I don't remember anything after the closing scene except for blurry vision and feelings of disorientation. I was crushed.

In the years of watching my cliffhanger version of the movie, I had always filled the gap of narrative with what I knew to be true. As a mixed-race child of a white mother and an Asian American father, all of my endings concluded with love. All of my endings suggested a future with kids like me.

I remember holding my mom's hand and stumbling dazed out of the theater. All I could see were people's legs walking in every which direction, and all I heard were voices buzzing, "Wasn't that great!" I wanted to yell, "NO! No, that wasn't great! That was horrible!" But I was too overwhelmed with tears. I was grieving.

When we got home, my mom asked why I was so sad, and I told her about all my imagined happy endings. She scooped me up in a giant hug. This was a defining moment of understanding who I was in a larger world. The reality of my loving mixed-race family was still a dream to so many others. Even the exaggerated world of a musical, with its over-the-top songs and choreography, its fictional characters and fantastical scenes—even that world could not render possible my existence. I would need to define myself. It would depend on me, and it would take a long journey of struggle against the external world, and a deep appreciation for how brave and strong my parents were in the face of it.

THE LESSON OF THE THREE-WHEEL TRACTOR

I am inspired when people figure out a very simple solution to a complex problem. I mature by making simple discoveries in everyday life.

There is a photo, slightly blurred, of a farmer and his tractor. It was taken in Europe, perhaps eastern Europe during the lush green spring. I can't tell exactly when the photo was taken because although the tractor is old—a 1940s or '50s model—it could have still been in use decades later. I know this because our tractors were also ancient. Farmers keep them like old work horses that never die.

The old farmer, in his fifties or sixties, sits on the tractor and is driving it down a street, on his way to get it repaired. It's obvious that there is a major problem: the front left wheel is missing. It's not a flat tire—the entire wheel is gone, the axle spindle exposed. Perhaps there was a farm accident, during which a wheel broke off and only the axle survived. Whatever happened, it was impossible to drive the tractor into town for repair.

I imagine this photo was taken in a poor region (although it could easily have been our farm or our neighbor's). I picture a farmer without a lot of resources. He couldn't bring a mechanic out to the farm; that would be too expensive, or one may not have been readily available, or they did not have the proper tools. But how do you get a three-wheel tractor to the repair shop in town?

The farmer probably didn't have a tractor trailer, or a truck big enough to pull a trailer weighed down with such heavy equipment. Plus, there was still the problem of the three-wheel tractor sitting broken in the barnyard or field; it would have been nearly impossible to load a three-wheel tractor onto a trailer in the first place.

So the farmer and tractor remained stuck. They had a problem with few options. I too would have felt stranded, trapped in a situation with few answers. No one had taught me how to drive a tractor missing a front wheel, and we didn't have the capital to own extra equipment like a truck strong enough to haul a tractor. I would have given up.

The old farmer in the photo must have experienced a similar moment of self-doubt. But he worked through it. The farm couple used their ingenuity and all the resources available to them. They needed to somehow raise the front of the tractor so it could be driven. They needed to find a way to balance their life.

The solution? The farm wife stands on the back of the tractor, opposite the missing front wheel. Her weight—she is a large lady—works as a counterbalance, lifting the tractor upright so it can be driven on only three wheels.

I admire this woman. She was valued for who she was. She knew she was not skinny, and she was proud of what she could do. She problem-solved with her partner and they worked as a team. They demonstrated the innovation of a typical farm family, where family really matters and everyone is put to work. They created their own rural innovation-training program: how to cope with limited resources.

This reminds me of the baling wire and, later, the duct tape solutions that literally anchored our farm. We lashed down aging wooden beds of wagons by tightly binding the planks to the metal base with wire. We reinforced an aging wooden support pole with multiple layers of wire wrappings. I myself am more of a duct tape problem solver; I cover tears, holes, and gaps with the silver magic, and my philosophy is that if a single wrap works, go ahead and add two or three more layers and don't worry about outward appearances. This method was especially effective during my teenage years, much to the chagrin of my father. Temporary fixes seemed appropriate, part of my future strategy of self-discovery: learning what needed replacing versus the things I could live with just the way they were (with the help of a little wire or tape).

When I look at the photo, I imagine that the old three-wheel-tractor farm couple approached their predicament with humor and goodwill. They seemed proud of their solution. They drove into town smiling and waving to all witnesses. They added a public face to the creative human capital that thrives in our rural communities. They displayed a balanced lifestyle and what it means to achieve success through teamwork and cooperation.

I love witnessing the low-tech and highly inventive. We often forget we have the creative capacity to discover everyday answers to everyday problems—especially when people understand their limitations and capitalize on their resources. I wonder who came up with the ultimate solution for the problem of the three-wheel tractor? In my imagination, it was the wife, who must have been an expert at creating balance between herself, her husband, and their farm.

I know of many men who would have tried using brute strength and equipment to muscle their way out of the situation. After initial failures, eventually a farming partner, often a woman, would quietly and patiently search for the solution, slowly positioning and repositioning herself and gently finding the sweet-spot solution.

To the untrained eye, this snapshot might look like a mockery of farm folk. But to the farmer, it speaks to the value of knowing yourself and the roles of others in this partnership we called family farming. The couple utilized their tactical advantage: a lightweight wife could not have worked in this situation. They also seemed to demonstrate positive body images—he too was not exactly slender. Together, they seemed to be a good match. No need for couples therapy. They appeared to love each other for who they were, not what they were not.

It did speak volumes about the importance of diversity, in this case with size. You could be large and strong—big need not imply unhealthy or lazy. And only with a type of body love could this couple publically demonstrate their imaginative solution to the question, How could you drive a three-wheel tractor?

In the end, this was an anthem to all farm women. Large and in charge. Big, bold, and beautiful. Every day was a love-your-body day. All images that took a while to be accepted in our farming communities. A world in which Nikiko will insist her voice is heard, demanding equality for all.

We could all discover something from the old farm couple. They seemed content and happy with themselves. We should all be so fortunate.

INVISIBLE FARMING

I was tired of being invisible. When I came back to the farm in 1980, few young people wanted to work in agriculture. Those from farm families had already left, running off to college to escape the dull, grueling work. City kids laughed at us "hicks" from the countryside; young adults from urban America never dreamed of a future in rural America. The media portrayed us as rural bumpkins living in hamlets with small-town mentalities and reactionary conservative politics. We were "slow," which implied we were dumb. I naively believed that since everyone ate, everyone felt a direct relationship with agriculture. Wrong. Worse than being forgotten, we were neglected. People needed food, not farmers.

Outsiders rarely asked about my life. They assumed I led a boring existence, working in the dirt, missing out on the excitement of city life. They imagined faded red barns, chickens as companions; I was a hayseed, nothing more. How interesting could it be to hear someone talk about watching grass grow? I wondered if I would impart this bitterness to the next farming generation, this part of the legacy of being invisible?

I began to purge the negatives—not the emotions but the people who made me feel them. I stopped associating with those who didn't give a damn about my work and life. I probably became a jerk too, selectively choosing my friends. I spent a lot of time by myself on the farm, where the orchards and vineyards harbored no judgments. They didn't inflate their egos, boasting of accomplishments while trying—consciously or not—to devalue my life choices. A hailstorm didn't discriminate between those with high- and low-paying jobs. A winter frost didn't care who you worked for or what title you were given. Nature doesn't discriminate if you graduated from an elite private college. We all suffered equally.

I hunkered down and surrounded myself with like-minded neighbors, but I also feared I had narrowed my focus too much. Had I closed myself to outsiders, circled the wagons, and adopted a them-against-us mentality? Maybe so, but at least I no longer felt marginalized.

It all changed in 1985. Peach prices were terrible, and our old heirloom varieties were suddenly blacklisted. No one wanted peaches that were not lipstick red and had a limited shelf life. No one hungered for old-style peaches with fantastic flavor. The market only sought cosmetically beautiful fruit; looks mattered, not taste. We sold thousands of twenty-three-pound boxes for fifty cents that season. Not fifty cents per pound but *per box*. We lost tens of thousands of dollars, and I knew we couldn't continue farming this way. I felt alone and isolated. I couldn't help but think over and over: my worth was valued at fifty cents a box—pennies per pound.

Depressed and rejected, a farmer couldn't do much at that moment, but a writer could. I wrote an op-ed for the *Los Angeles Times* entitled "Epitaph for a Peach," describing my plight in a sad world where flavor is lost. The *Times* syndicated the essay on its wire service and the article was picked up nationwide.

I then began to receive old-fashioned handwritten and typed letters, initially one or two a day, about thirty to forty letters total.

Random readers urged me to "keep something with flavor" and "save a peach with value." Their support overwhelmed me. Armed with this handful of letters, I approached Marcy: "We just lost tons of money, but look, I got these letters of support."

Wonderful letters of encouragement.

I looked directly at my poor farm wife, who had just returned from her work *off* the farm—a job that paid our expenses and gave us health insurance benefits. "So which is more important?" I pled, meaning money versus the values expressed in the letters.

She immediately knew the answer. She understood the value of her career. We both smiled weakly.

The phrase "artisan peaches and nectarines" had a nice ring to it when spoken aloud. The economics of small-scale enterprises rarely made sense anyway, so I redefined myself as more of an artist than a businessperson: always on the outside looking in, a misfit but at least a contently creative one. Of course we needed to earn enough to keep living—I joked to friends that I wanted to pay income taxes because it implied we had made some money that year—and Marcy's work as an educator could support my farming habit. But we were also making new social connections, and maybe, just maybe, there was a community out there who'd support us as well.

Our farm operated outside of typical reward systems. We would stay strangers to huge piles of money. Traditional symbols of success would elude us. Our only validation may lie in the future, when we can look back at the rearview mirror of life and realize we did some good, growing peaches with flavor for a small but appreciative audience who still understood what real taste was all about. Perhaps our children would inherit more than debt.

We grew something of value while keeping and sharing our values: our marginalized, almost secret life of a farm family, and a humility that's often hidden by and lost in the noise of modern existence. Perhaps our goal was to be recognized for something that was so simple it appeared to most outsiders as nothing at all: a transparent life built on a network of relationships and on sharing the meaning of emptiness, like that feeling in the stillness of a summer morning just before the rush of harvest, or in the solitude of standing on the farmhouse porch in the middle of the night, thinking, reflecting alone.

In those moments, even as I confronted the struggle of passing down to another generation the curse of being invisible, I found the freedom and strength to move ahead.

HOW TO FRENCH PLOW

Spring is French plowing season, when farmers break the earth's winter crust, and the smell of freshly turned soil fills the air. I learned how to use this tool from my father, who learned from his father. We have three or four generations of this tool lying around on the farm, collected over the years and left discarded in our junk pile as memorials, but now Nikiko is learning how to master this implement. To farm organically, without herbicides, we have resurrected this old weed-control tool from the past.

For vineyards, the French plow is the tool of choice to mechanically dig weeds out from under grapevines. The plow is pulled behind a tractor and accomplishes the tricky task that challenges grape farmers: how to employ a cutting blade between two vines without damaging the vines themselves. The plow must first swing past the trunk, then ease back to till directly under the vine trellis and canes, and then swing back out again to dodge the next vine. An 1878 report from the Universal Exposition at Paris first described a crude plow the French used to cultivate their rich vineyards and "keep the soil open and free from weeds," and eventually the implement was imported, copied, and produced in America and transported to California vineyards. (I was shocked to discover that Americans credited the French for this invention instead of appropriating it for themselves.)

Over time, the device was modified and redesigned, but that didn't necessarily make it any easier to use, at least not for me. At our farm, many vines sacrificed themselves in the name of "progress," or that's what, as a young man, I had told my father following my latest French plow accident, in which I'd hooked a vine and ripped it right out of the earth. Not many farmers use the French plow anymore; modern chemicals and machinery have replaced the backbreaking work.

While reliable, a French plow requires a farmer to drive slowly—really slowly. I've spent hours in my vineyards, going up one side of the vine row, then down the other, the tractor crawling along, the plow digging weeds around a single vine before moving to the next. My back and neck ached from driving forward while constantly looking backward to monitor the plow behind the tractor. It's hard work, but mastering that particular point of view is essential: great peripheral vision remains a skill each farm generation passes down to the next. One false move and the blade will hook a vine—no "do-overs" with the French plow. By the end of the day, I would see progress, but I would also feel disheartened looking at the remaining rows still to be plowed. The weeds would sneer and grow taller in the warm spring air.

Nowadays, I have it lucky. My plow is equipped with modern hydraulics that automatically pull the blade away from the vine. Our old French plow rests silently in the junk pile with its ancestors. One, now mostly just pieces, is from around the time tractors replaced farm horses and mules. Dad once showed me how to maneuver it.

As late as the 1950s, he still employed farm animals, and his two mules, Jackie and Molly, pulled a French plow that sliced into the earth with a metal blade (some very early ones were wooden).

Walking behind the team and implement, Dad tightly gripped the plow's handles. As a vine approached, he had to lean left to twist the blade, swing the plow around the vine trunk, then quickly lean to the right to make the blade slice back under the vine trellis. I'm sure this worked better in theory and description than in actual practice. "It had a certain rhythm," Dad once explained. I believed him, but it also had to be hard, hard labor that would break down a farmer's body over the years.

The next innovation was the addition of a retractable cutting blade. It scraped against the vine trunk and was pushed out of the way by the impact, then slid back in place after safely passing the trunk. A farmer had to stake his vineyard and reinforce weaker vines to withstand the pressure of the blade. I imagine that during the first season using this new and improved version, a lot of the older or weaker vines did not stand up to the tension and were simply shoved out of the way and cast aside.

Then, with the introduction of tractors, the driver (as opposed to the "wrestler," which was the term my dad used to describe the farmer controlling the early horse-drawn models) worked the implement, guiding it away from the vines. A kick-arm was added—a type of trigger that helped sense when a trunk was approaching and then engaged a spring to push and pull the plow in and out of the vines. Tractors can't, of course, feel when the plow hooks a vine root. A mule, according to Dad, knows better and simply stops, not wanting to struggle any more than necessary. Tractors don't have that kind of common sense.

French plowing means learning to live with accidents that last a lifetime. I can still identify where I've hooked vines yet reacted quickly enough to clutch and brake, only moving the vine a foot or so. The roots held fast, though they stretched a bit, and from that point onward, the crooked vine trunk will be forever associated with my mistake, as it should. Other times—and this happened often—no matter how rapid my reflexes, I would accidentally rip out the entire vine from the earth. The vine died. I felt bad. My father mourned. Every time I would drive past the gap (or, worse, Dad did), the empty spot memorialized the incident.

To this day I still use a French plow. It has its benefits, including the ability to slice surface roots, which encourages the vines to become more deeply rooted. I also believe it is healthy for soil to be turned, mixed with green grasses and weeds in order to build soil life churning with microbes. Since I farm organically and thus can't use conventional herbicides, I don't have many options other than mechanical weed control.

I am forced to live with my weeds and now know them by name. Some have escaped the wrath of the blade, survived, and I have since discovered they are relatively harmless, like chickweed and purslane. Others, such as Johnson grass, are downright evil and require deep plowing in an effort to dig out as many of the roots as possible. Sometimes weeds that seem innocent in the spring, such as mare's tail, trick me into ignoring them until it is too late and they are poking out of the grape canopies, stealing sunlight.

French plowing links me with the past. It is not nostalgia, but a real connection with my father and with a sense of farming history. Each plowing season, I learn a little more about physical work and appreciate even more what he accomplished in his time, with his tools. I smile every year when I finish French plowing. I look back at the plowed fields, a moment shared with generations on this land. I believe that the best and simplest advice for my and the next generation's journey into farming came from my father. He would suggest that when you French plow, "simply go slower."

THE SCENT OF RAISINS

Around Fresno, at the end of August and in early September, the scent of drying raisins fills the countryside. In the evening, I drive slowly with my car windows down and I can feel the subtle caramel fragrance of grapes curing in the sun. Yes, I believe you can actually *feel* this if you understand how raisins are made.

It goes like this: you work the entire year, then in the late summer you pick green grapes, lay them out in the sun to dry for three weeks, and if you don't get rain, they are transformed into raisins. It sounds wonderfully simple, and it is. And also insanely stupid. We expose our harvests to nature and the elements, the process is extremely labor intensive, and it doesn't always make economic sense. Yet people who grow raisins are nothing if not persistent. If the average age of a farmer in the United States is fifty-five, add about a decade or two to us stubborn old raisin farmers. We are old, hanging on to a dying tradition, and we will probably pass away still full of optimism, hoping nature will cooperate one more time and we'll get our raisin crop in with just enough reward and profit to do it again and again. I suppose we must enjoy pain.

Raisins have been around for centuries. Armenians were said to have discovered the art of drying grapes in the sun to make raisins, and the tradition lives on in the large Armenian community around Fresno. Some of the major raisin-processing plants were owned and operated by California Armenians.

The first raisins in California were produced in the early 1900s, and immigrants in particular were attracted to this labor-intensive cash crop. Among the groups that embraced the raisin business were Armenians, Italians, Japanese, and Latinos. Even today, ethnic diversity thrives in the raisin community, which now also includes Asian Indians.

Raisins are made by the sun and not in a factory or processing plant. (Note: I'm talking about traditional raisins here, not the golden raisins that are often used in baking, typically for their color rather than their flavor. "Goldens" are highly processed and artificially dried in tunnels with sulfur and heat. Many incorrectly believe they come from green grapes whereas dark raisins come from dark grapes; a golden raisin simply never gets the "suntan" that gives its counterpart that dark tone.) One method of making raisins—the more traditional way—begins in August, when the grapevines are watered for the final time and the earth between the rows is disked free of weeds. The dirt bakes in the heat and dries into a powder before it is terraced with a gentle southern slope. Then green grapes are picked by hand and spread onto paper trays, each sheet about two by three feet. One acre produces on average a thousand of these trays. (Even on our small farm we've had fifty thousand trays a year.) The grapes are left exposed to the sun for weeks, curing into raisins, which are then loaded into wooden bins and sent to processing plants to be cleaned and packaged.

New changes have recently been introduced to the industry, including a method of drying grapes on the vine and then harvesting them mechanically. This saves labor costs and can be accomplished on a larger scale. My farm, with its small acreage, is quickly becoming a dinosaur operation; we make about forty acres of raisins, still using trays, still employing numerous workers to harvest the grapes by hand. Over the next decade, hand picking will give way to machines, and our farm will be forced to change.

Every harvest is filled with uncertainty, and yet that circumstance used to be perfect for struggling immigrants who were willing to take risks and dream of big profits. Raisin farm families all have stories of watching rain clouds march into the valley drenching green grapes trying to dry. My first years back on the farm after college, we lost our entire crop twice in three years. It rained, mold grew on the half-grape/half-raisins, and my father silently hooked up the disk to our tractor and plowed under the entire field. Gone in a few hours. The pain still lingers. Welcome to farming.

I have heard stories about a farm couple, just a husband and wife, together carrying and stacking two-hundred-pound "sweat boxes" of raisins at harvest time. The wife said, "Oh, I didn't know how we lifted those wooden boxes, but we did."

I knew of farm families and farmworker families laboring together, kids and adults, racing to save a crop, rushing against the approaching storm to roll up trays and protect them from the rain. For a moment the lines between owners and workers blurred. All parties recognized what was at stake.

Am I nostalgic for the old days? No, things do and should change. Yet I am uncertain which traditions need to survive and which antiquated methods will retain value into the future. What I do know, though, is that during raisin season, my senses are piqued and fully engaged. I can determine the moisture of almost-ready raisins by the feel of them in my hand. The scent of rain in the air sends a panic through my system. I feel connected to the harvest.

We have become more of an industry now, less stories and more business. Machines don't carry the same emotional attachment to the land that workers do. The hope of scratching out a living in order to buy a place of your own and then send your kids off to college someday has been lost in the economics of modern-day farming.

But, for a few more years, I will make raisins the old-fashioned way. Younger generations will shake their heads in amazement at the hard, hard work of their immigrant great-grandparents and question the return on investment. I'll take a deep breath of the dusty and caramel aromas and simply enjoy the moment. Before we know it, the scent of drying raisins will be gone and we'll have just the stories and memories.

REVOLUTION ON THE FARM

On one wall of our old red barn hangs a "singletree"— a horizontal wooden bar with a metal ring attached in the middle and fasteners at the ends. Farmers once used this to hook up a team of horses needed to pull a farm implement or wagon or trailer. I have no clue how to use it; I have never farmed using a horse or mule. My tools involve gas and diesel.

I remember my father talking about farming with animals. He was a poor, young farmer in 1946 and had no choice; he couldn't afford the new technology. My father didn't know it then, but that year marked a transformation in American agriculture: the moment tractors overtook horses on farms, the moment innovation trumped tradition. If my father farmed with horses and I farm with tractors, what will be the revolution for Nikiko? What else will be displaced in the name of efficiency?

Early in the 1900s, Henry Ford first introduced the Fordson, a modern tractor with a small gas-powered engine. The Model F was farmer friendly—if the average farmer were willing to tolerate the unbearable engine heat and a motor that occasionally caught on fire.

Imagine millions of farmers looking at this new invention and shaking their heads, saying, "Ain't gonna work." They then returned to their farm and their loyal and steady team of horses or mules. They could not foresee the changes that were coming, whether they liked it or not. They didn't understand the thinking behind the arrogant yet visionary quote often attributed to Ford: "If I had asked people what they wanted, they would have said faster horses."

Gradually, the Fordson improved. It still required inputs of gas and oil, and used up resources of steel and, later, rubber, but it could outperform the poor old horse.

The tractor triggered a massive change in farm size. In 1900, farmers typically worked about one hundred acres; by 1960 it was three hundred acres, and today it's more than four hundred acres. Some claim the push for productivity forced farmers to keep expanding in a never-ending race.

Greater productivity in agriculture then resulted in cheaper and cheaper food; the efficiencies of the tractor and other innovations made farm products more widely available than ever before by reducing prices. According to the Bureau of Labor Statistics, the percentage of annual income the average American spends on food has continued to plummet. In 1900, when horses ruled our farmlands, Americans spent 43 percent of their income on food; by 1950 it had dropped to 30 percent, and today it's less than 9 percent. Tractors ruled, horses drooled.

American farms also changed in another significant way: human labor was no longer needed as much, and the farm population dramatically shifted. In 1900, 41 percent of Americans claimed to be farmers and 60 percent of Americans lived in rural areas. By 1960, when tractors dominated agriculture, the number of farmers had collapsed to only 8 percent, and rural communities shrank to 30 percent. Today, less than 2 percent of Americans identify as farmers, and rural residents comprise only about 20 percent of the population.

Were millions of workers displaced or freed? Other than periods during the Great Depression, those people migrating off the farm were not necessarily economic refugees out of *The Grapes of Wrath*—the unemployed poor fleeing the terror of the tractors pushing them off the land. For many, the tractor offered a welcome escape from much of the hard physical work of farming.

Industrial America grew at precisely this moment in our nation's history and absorbed this new, massive labor force. Commerce benefitted from a fresh set of skilled hands, and this launched the rise of the urban middle class. The tractor, far from being a mechanical demon, was seen as a savior, freeing millions from the torture of farm labor and isolated rural life.

The American tractor: the perfect poster image of capitalism's creative destruction. Innovation inevitably destroys old systems. Progress? Not if you were a horse.

But what are the indirect costs of cheap food? Major medical problems are clearly associated with access to the inexpensive processed food with which Americans stuff themselves. Indirectly, are obesity and diabetes connected with the tractor and the rise of mass-produced commodities in search of a mouth to feed? Or is this all a moot point, long after the horse left the barn, since farmers will never forsake tractors.

Are bigger farms necessarily better? Some claim that the accumulating environmental bills have yet to be paid by large-scale agriculture. In our valley, some communities have had to stop drinking water from their wells, as excess nitrates, often the result of modern agricultural methods, have polluted groundwater supplies.

To compound the issue, fewer people on the farm mean fewer voters supporting healthy agricultural policies; we forget that tractors have also displaced rural political clout. The farm vote is now so small it simply doesn't count. More and more, policy decisions for agriculture are determined by "city folks." It is not about politics, it is about geography.

We have a number of tractors on our farm. They don't require daily feeding, and they can work for hours and hours pulling heavy equipment through my fields with ease. Daily, my body thanks the arrival of tractors. But tractors are not alive, and I find little emotional comfort in their cold metal and the smell of diesel. So when I notice the horse tack hanging in my old barn, I see more than the rise of tractor power. I see the consequences of innovation: with succession comes displacement.

I shared this story with Nikiko, and she had trouble getting past the idea of having horses on the farm. I then realized that I sounded very old.

NIKIKO'S FIELD NOTE: CULTIVATING TRUST

Only once in my lifetime have I participated in planting an orchard. Our farm bucks the trend of changing orchards every decade, and instead we try to keep our orchards healthy and regenerating for as long as possible. (Our oldest orchard is now close to fifty.) When I returned to the farm after undergrad, however, I witnessed my first planting: we were preparing a new orchard of nectarines, Rose Diamond.

We started by bulldozing half of the Spring Lady orchard. These were trees planted the year I was born—they were intended as my college fund and had fulfilled their purpose—and when I came home, degree in hand, the time was ripe for change.

One of my jobs was to help prepare the earth after the old trees were removed. We disked the soil, churning the dirt and leaving trails like a comb—our own huge-scale Buddhist sand garden.

My dad knew there were a few patches of hardpan still underneath the surface, and he tasked me with trying to break them. To the back of our four-wheel-drive tractor we attached a piece of equipment I had never seen before: a contraption with two three-foot-long metal teeth, both hooked slightly at the end. When submerged in the soft dirt, these teeth moved through the ground, grabbing, pulling, and breaking obstacles like old roots.

I worked the ground with this new-to-me tool, feeling the teeth glide until I hit something. The teeth hooked underneath a sheet of hardpan, and the once-powerful tractor started to struggle. I clutched, reversed, and went forward again, hoping a little more speed would give me the upper hand. No luck. After a bit more pulling, a piece of the hardpan finally broke off and, with a jolt, the metal teeth were free again. I had taken off merely a corner of the hardpan; a much larger sheet still remained.

We disked more, and slowly chunks of hardpan broke and came to the surface like debris floating in water. Amazing. I got off the tractor and marveled at the formidable teeth of my metal companion.

After the saplings were in the ground, my dad and I walked every row and greeted each tree with water. At one point he slowed down and asked me what I thought. I said, "Well, it's exciting." He replied, "Yes, when you're forty these will be full grown." I stopped in my tracks. That was twenty years from now! I hadn't ever stopped to envision my life in twenty years, but here, in the fields, my future was already being mapped, and I sensed this transition would also bring me face-to-face with aging: both my own and that of my father.

My dad would not always be there. Every year on the farm now became a precious opportunity to learn how to go on without him by my side. How many winters did I have to master pruning under my dad's tutelage? How many new orchards would we plant together before *I* had to know who to call, how to measure rows, how to plan for the future?

My first few years farming, I scurried to write down these little droplets of knowledge, whether they came at the dinner table or over greasy gears and dust as my dad and I fixed an old piece of equipment. Three years into my journey, I had a binder stuffed with disorganized scribbles on Post-its, napkins, torn pieces of paper, and a half-written farm blog. (I had once set out to write each day about what I had learned on the farm, but I failed to keep it up.) These scraps of farm journals are extremely useful, but keeping consistent notes has been challenging, even as a person who is otherwise extremely disciplined. Knowing my time to learn is limited only adds to my frustration.

Yet the notes are valuable, and I continue to keep them as well as I can. They have saved me when my dad was away at a meeting and I had to remember how to French plow, or when something broke and I was alone in the middle of a field. And these notes aren't just for beginning farmers like me. My dad constantly refers to the records of what he has done in previous years. Since our perennial crops require that sometimes we only practice a job once a year, his notes are his pedagogical tools too, even after decades of use.

But the notes are missing something. They can't quite capture the arc of an entire year. Cause and effect are sometimes mysteries we must wade through with guesses. (Why were the fruit so small last year? Was it weather? Fertilizer? How we thinned? A combination?) And of course even the most thorough notes do not help me work through the feeling that I'm preparing to farm without my dad.

It will be crucial for me to remember one day that 2013 and 2014 were drought years and to know at what level our groundwater stood then. How we fared, what happened to the trees—those experiences will be important to archive in farm journals.

But learning how to farm alone...that's a test that I'm struggling to accept, knowing that my farm journals will never be able to replace my dad. One day, my notes may help me remember, like my dad does now, which rows are harder to irrigate, which tree in an orchard suffers from aphids, and maybe I'll recall—without my notes—where I was when I ran into hardpan while planting a new orchard. For now, I struggle to ease into this reality. I tell myself to breathe in the discomfort of uncertainty; I tell myself it's not healthy to think about death so much. I think of my dad reading this and saying, "Gee, I'm not dead yet!"

Maybe behind the utility of the farm journal there is something else: a symbol of the challenge of growing up. In it, I record what I've learned so I can teach myself again in the future. I must cultivate trust with myself, a belief that I can do this, that I will know what I need to know when the time comes. I hope that this confidence, plus my farm journals, will be enough to carry me through the fear of one day doing this alone.

A FATHER'S STROKE

I'd broken the French plow, driving too fast, always in a rush to be more productive. The plow had pushed against a vine trunk, and a deep-seated root had hooked my blade. Not only had a metal brace bent and snapped but, worst of all, I'd simultaneously pulled half the vine out from the earth, dislodging decades of growth. I tried to push the trunk back into the dirt and cover the exposed roots with soil. If I were lucky, the plant would reestablish itself following the abrupt uprooting. I hoped for a transformation, the vine accepting its altered place in the landscape.

Tractor and I stumbled home to the shed. The wounded attachment demanded repair. An acetylene torch could heat the bent metal so I could straighten it, and a new bracket could be welded on to strengthen the old. Actually, my father did most of the welding, restoring the damaged apparatus to save money, using his savvy to renew the old in order to keep expenses low. We rarely purchased anything new; used equipment was his secret to compete.

At seventy-five, my father was still much more adept than I with the arc welder we used to mend, join, and reinforce steel. And at forty-three, I still happily accepted his paternal help. I broke things; my father fixed them.

As I drove up to our shed, I saw my father wandering around his tractor. The engine was roaring. Dad shuffled in the dirt, poked his head near the engine, paused, pushed his cap back and scratched his head, then walked around the front and repeated the motions on the other side. Standing next to him, I leaned in to listen too; I thought he was trying to detect a problem from the sound, using all his senses to diagnose a problem.

Then I glanced over at him and was shocked by what I saw. His face was drooping, the right side having collapsed into a deep frown. He tried to speak, but only mumbled sounds escaped, along with drool from the corner of his mouth. He rubbed his temple, grimacing as he tried to massage out the pain from inside his head. He was suffering a stroke.

I reached for him, pulling his right arm over my shoulder, trying to lead him to the house, but he jerked us back, refusing to move forward. He pointed to the tractor. I leaned over and shut it off. Dad still fought me, reaching for something. His hand pointed to the seat cushion. As the final act of a workday on the tractor, we traditionally flipped the cushion over, which kept the morning dew off the surface for the next driver the following day. Dad insisted we complete this act before staggering away.

Days passed. At the hospital, Dad was still alive but in a coma, which the doctors said was not good. "He should have woken up by now," they reported. I felt helpless, waiting for a sign, but I could do little sitting next to his limp body in the hospital bed. Farm work still needed to be done; although I was worried about my dad, I welcomed a return to the fields.

His tractor was in the exact spot we'd left it. It was parked at an odd angle, and I could still see the tire tracks he had made in the soft dirt. As soon as the stroke had hit, my father had raced in from the fields, weaving and wavering on the tractor as he fought for control, the pain in his head intensifying.

I took a deep breath. My father was still with us, barely. I didn't want to think this was where we took our last steps together. I blinked back tears.

I removed the key from the ignition and opened the metal toolbox mounted on the side of the tractor. Inside, I saw an orange, remnants of my father's last meal on the farm. He must have planned on having a snack in the fields instead of driving back to the house for a break. (The farmer ethos considered such interruptions a waste of time.) He had planned to park in the shade of a peach tree, peel his treat, and enjoy the flavors from our backyard. During such respites, he would have studied the shoots emerging from the branches, dreamt of a productive year, and anticipated another season of hard work. A new year had begun. One more season to scratch out a living in the dirt. Another cycle of life.

The orange evoked the image of Dad's taking refuge in the fields, enjoying the occasional pause, a snack of refreshment from his own land. The fruits of his life's work. Each bite must have been the sweetest. I held the orange in my hands, running my fingers over the skin. I couldn't help but notice the thrip scars. This was an organically grown fruit; on our farm, we tolerated these imperfections. They were all part of the natural system at work.

I felt like we would never be whole again. Even if Dad did awaken, I could only hope enough of him would recover so that I could carve a work place for him. He would suffer, knowing that he would never be the same. And I knew I could never be as dedicated as he had been. The farm would never be the same.

Dad was still hospitalized, but I had to continue with chores. Nature didn't stop for personal reasons. The land needed to be plowed. Weeds flourished with the warmth, and we brought the shovels out of hibernation. This was the season for work gloves.

Walking through Dad's shed proved difficult. His memory lived in his workbench, his hand tools, his old pickup. I found a pair of gloves standing upright, stiff from his sweat, the fingertips curled as if still clutching a tool. The old leather seemed to be reaching out, trying to grasp something.

In a moment of foolishness, I tried to pull them on. They didn't fit. They were molded to hands of the previous owner. I could never fill that space.

Ghosts lingered on this farm, the spirits of those who once worked this land my constant companions. Generations had left behind their pruning scars on the vines and trees. Still piled against the ditch bank were mounds of hardpan rock that Dad had once removed from our fields—the very rocks that had made the property cheap enough so that a struggling immigrant family could afford to buy it and plant roots. Even the weeds, which thrived as a result of my organic farming methods, served as my contribution to the reservoir of decaying matter that created healthy soils to feed our luscious fruits.

Even if I could never fill my father's work gloves, I knew these stories would always surrounded me and I would never farm alone.

Dad woke up a week after the stroke, the old farmer opening his eyes with family as his witness. I was still in the fields. I had just finished the day's work when I got the phone call that something amazing was happening. I wept in the shower before rushing to the hospital.

The right side of his body was damaged and required months of therapy. We began to use the term "insulted" instead of saying he was "injured" by the stroke. I liked that term, figuring a word that triggered an emotional response would bolster his recovery.

The stroke had insulted his body, and we would respond to the insult.

One reality settled in quickly: I could no longer depend on my father to fix things. I was forced to practice arc welding, the technique in which metals are bonded together using heat generated by an electric current. From an electrode held only a quarter of an inch away from the object to be welded comes an electrical arc that liquefies the metals so they puddle before then quickly cooling, recast as a solid joint, if done correctly.

My own welds looked sloppy and weak. I practiced and practiced, but the seams were never as good as my father's. Mine looked like the joints of a blind welder. And perhaps I was. Dad had used an old-school welding helmet. The eye slit was covered by a very, very shaded lens, black enough to protect his eyes. The problem was that the dark lens was so dark you couldn't see much until the arc was created. I had trouble getting the damned arc to ignite and glow. So I remained blind.

Eventually I gave up. I visited a welding store and discovered a new generation of helmets with lenses that automatically darkened in response to a glowing arc and that used tinted filters to protect the welder's eyes. I could finally see and started to fix a few things without my father.

After a long day of repair work, I cleaned up the area and stood before the wall where Dad had hung his welding helmet. I paused, unsure of where to place the new one.

PART TWO

HOW WE LEARN: PRESENT TOOLS

IT'S SO HOT THAT...

During the summer, our valley is hot. You can't escape the temperatures, no matter what you do. Nikiko grew up in this heat, so she's no stranger to sweating. I can't recall her whining about it as a child, but perhaps she simply didn't know better. She is a child of heat, and over time I have convinced her that the peaches and nectarines love this climate and therefore so should she. People in farm families are so often naive in this way. We have to be.

"Peaches originate in the high deserts of China," I explained when Nikiko was in middle school. We stood in our farm shed, packing peaches for upscale restaurants. I tried to convey our quest for the perfect peach and why we took great pains to grow, pick, and pack only our finest. It was 110 degrees. We then each bit into our own soft, gushy, slightly overripe peach, and they were warm inside.

"It's like eating a cobbler without the dough," I said.

"Add heat to peaches and it brings out the flavor," she added.

Nice observation. I knew then she'd make a great cook.

We struggle, though, with heat and sweat and juicy peaches. The fruits can go soft if we delay; a few hours can make a huge difference. But sometimes heat can actually slow the ripening process. If the trees sense it's too damned hot, they shut down. The peaches hang, almost ready but not quite. It's as if the trees have decided to take afternoons off, a natural peach siesta. I can practically hear them proclaim, "Crazy to be working in this heat!" They're much smarter than the farmer.

The best lesson is to teach Nikiko how to joke about this heat.

Twenty consecutive days over 100 degrees, and some over 110, may not be a laughing matter, but you can't control the weather. It's often still over 90 when the eleven o'clock nightly news comes on. There's nothing to do but accept it. Welcome to our farm.

But we can both acknowledge this grim reality and turn it into a positive experience if we can learn to laugh at ourselves. Make lemonade from lemons...well, at least hot lemonade. And laugh to make it through the heat wave.

Over the years, I've collected some well-traveled one-liners, and every year, I hope we can survive our summer weather with a little comic relief and some light-hearted humor. Apologies to those who believe they originally coined these jokes, but I believe the first Native American who set foot in the Central Valley in the middle of summer thought of most of these.

It's so hot that...
Even the sun is looking for some shade!

"But it's a dry heat," a friend rationalized about Fresno and our valley. Yeah, well, so is your oven, but you don't stick your head in it.

It's so hot that...
Only in our valley can you break a sweat the instant you step outside at 7:30 a.m. before work.

You can't take a cold shower because the pipes are so hot. And the exposed pipes are *really* hot.

It's so hot that...

They're considering changing the name from "Valley of the Sun" to "Surface of the Sun."

During our prolonged drought, a sad valley resident once prayed, "I wish it would rain—not so much for me, 'cuz I've seen it—but for my five-year-old."

You can measure highs with an oven thermometer.

It's so hot that...

You burn your hand opening the car door.

You learn that a seat belt makes a pretty good branding iron.

The AAA Rescue Team had to "unstick" you from your vinyl car seats.

You discover that in July it takes only two fingers to drive your car.

You find out too late that you can get a sunburn through your car window.

It's so hot that...

The best parking place is determined by shade instead of distance.

Your car overheats before you drive it.

You realize that asphalt has a liquid state.

Driving distances are measured by bottles of water rather than tanks of gas. You calculate travel according to your MPB—miles per bottle.

It's so hot that...

You can make instant sun tea on the front porch.

Even the jalapeño peppers in the garden are hunting for shade.

Potatoes cook underground, and all you have to do for lunch is pull one out and add butter, salt, and pepper.

You buy a loaf of bread, but by the time you get home it's toast.

Not only can you fry an egg on the sidewalk, you can cook hash browns to go with it.

You eat hot chilies to cool down your mouth.

It's so hot that...

Farmers are feeding their chickens crushed ice to keep them from laying hard-boiled eggs.

The cows are giving evaporated milk.

All the corn on the stalks started popping and flying through the air. The cows thought it was snowing.

You saw a bee take its yellow jacket off.

It's so hot that...

Anyone who works in the heat shares a common bond—farmers and construction workers become brothers and sisters.

Outdoor work still has to get done, so people start as early as possible, sometimes by 5:00 a.m., pretending they can see in the dark.

You drink gallons of liquids but still cramp from lack of fluids.

Astute sports teams bring bananas for time-outs.

It's so hot that...

Appetites may decline, but we fill ourselves with empty calories and are willing to drink most anything with ice in it.

Meals become a challenge—no one wants to bake in this heat—so some will go out to eat, taking malicious joy knowing someone else is slaving over a hot stove.

It's so hot that...

You would give anything to be able to splash cold water on your face.

The bottoms of your tennis shoes feel like they are melting.

You can attend any function wearing shorts and a tank top.

It's so hot that...

All the water buffalo at the Fresno Zoo have evaporated.

The trees are whistling for the dogs.

The sparrows have to pick up the worms with potholders.

The doves are laying their eggs sunny-side up.

Lizards carry leaves to put under their tails when they sit down.

You saw a dog chasing a cat and they were both walking.

It's so hot that...

When the temperature drops below 95, you feel a bit chilly.

Hot water now comes out of both taps.

Friends back in Wisconsin ask why you want to live someplace so hot, and all you can say is, "You don't have to shovel hot air!"

Your biggest bicycling or jogging fear is, "What if I get knocked down and end up lying on the pavement and cook to death?"

Your lawn has burnt, so you don't have to spend the time mowing it. But the dead grass makes you feel even hotter every time you look at it. And it may be a fire danger.

It's so hot that...

A friend coming to visit asked if he should bring suntan lotion. You told him, "No, just bring your dental records."

The old pick-up line "you're so hot that when I look at you I get a tan" is misinterpreted in hundred-degree heat.

Someone told you to go to Hell, and you thanked him for the suggestion of how to cool off.

And finally, it's so hot that...
You can say "113 degrees" without fainting.

To the next generation of people who will work these lands, I say welcome to the dog days of summer here in our valley, our home. Hopefully, a joke and a smile will make you feel a little cooler, or at least forget your misery for a few moments. And I wonder if, in the future, Nikiko will have new "it's so hot" jokes of her own. Or can you still recycle these?

Either way, I have concluded that if we share some humor with each other, we can blow off some hot air and feel better inside.

RITES OF SPRING

Spring arrives with the first warm breezes and fogless mornings in our valley. I disk the soil, breaking winter's hold on the earth. The peach and nectarine trees awaken with blossoms, first revealing their pink buds before blooming into a glorious canopy. Millions of blush-colored dots blanket the landscape. A new year has begun.

But this year, harsh memories of the cold, bitter winter linger because it rarely rained. Welcome to the new climate age, where massive swings in weather have become the rule.

Every spring, I plow the earth and something is plowed into me. Usually it's the spirit of the land, a sense of renewal, a bonding of family with the earth—now including our daughter, who has come back home to work the farm. This year, however, I begin work with a new realization: change, especially that of the weather, is the new normal.

The lack of rain troubles me the most. We'll get very little surface water, due to a limited snowpack in the Sierra. I can pump from my wells, but water tables can quickly drop, and wells can go dry. Most of the Central Valley has received less than half of its normal rainfall. This could change with a late March miracle, but long-range forecasts are not optimistic. The traditional method of coping with change is to simply work harder. Determination. *Gaman* is Japanese for "perseverance." *Gambatte*, Japanese for "endure."

Of course, what is normal? Typical weather models are based on thirty-year increments, counted by decades. So if you were born in the 1960s or earlier, your weather memories don't count (lending credence to my claim that as I get older, the weather just isn't what it used to be).

Droughts are common. In addition to the historic drought that began in 2012, in just the past forty years we've also had them in 1976–77, 1987–92, and 2007–09. We farmers live with this risk, and lack of rain has been fairly common in the last century. Old-timers who remember the Great Depression will recall California's decade-long drought beginning in 1928.

But we have been spoiled; we're ending a century of abnormally consistent weather. We developed farming systems built on a culture of expectation. When we look at much longer timelines, the relatively wet period of the twentieth century has been the exception rather than the rule in California. Our few dry years have typically been followed by extremely wet rainy seasons.

We're entering a new "weirding time." Much more volatility is to be expected, with extremes in weather part of the new norm. There's still debate concerning how much of the difference is a direct result of global warming, but it's clear: something is changing. We have to find a way to live with change; discomfort may be the new normal.

Across the nation, a warming trend is advancing northward. Planting guides now suggest gardeners can experiment with species typically grown in more southerly regions. The USDA updated its climate zone guide to accommodate the rise in temperatures and shifting of boundaries. That doesn't mean I'll plant bananas and pineapples on my farm, but it may mean spring will come sooner each year and weather patterns will be more erratic.

Change comes slow to a farm and its farmers. But I know that a few years of severe drought demands immediate actions.

In San Diego, some farmers are "stumping" their avocado trees, literally cutting them in half at the trunk, to save water; they will lose a few years of crop, but it's one way to try to keep the investment alive. Other farmers are switching from lower-value row crops, like vegetables and other annuals, to higher-value tree crops, hoping to earn more from their limited supply of water. These farmers will quickly learn a new reality, however: perennial crops can't be uprooted and transplanted; you have to plant with the future in mind.

Even on our small farm, we've begun a gradual process of change. One of the smartest acts I did years ago was to fallow fifteen acres, much to the chagrin of my father. He grew up with the premise that you farm every precious acre you have. Nikiko will have to live with my decision, and I hope it meets with her approval long after I'm gone.

I rationalized my decision to bulldoze an idle part of our farm: Why fight these swings of weather, not to mention the poor prices? (This was a decade ago and I was talking about raisins.) With a new pioneer spirit, I pulled out old vines and, among other benefits, created a new avenue that now splits the farm. What a joy every spring as I rediscover the new shortcut on the farm. Who wouldn't jump at such an opportunity in life?

Major shifts in weather point to a new challenge: survival in agriculture will be based on the ability to change. I imagine it as a two-tier strategy. One is based on the very large model—economies of scale benefiting the biggest and most efficient operations—and the other works for us small operators who are adept at the culture of change; we accept, adapt, and adopt while finding our niche in the new food chain.

What I'm not so confident about is policy and technology. We don't yet have policies in place that are equipped to cope with this new normal. For example, we are still fighting over water as if we're clans locked in tribal warfare. We cling to a myopic sense of time that spans only a few years at the most. Once, farmers had long memories, but the current business of farming demands historical amnesia; measuring the return on our investments means thinking in quarters and years, not decades. But what happens with a thirty-year drought?

Many believe that we can invent our way out of these problems. Technology has created miracles, productivity has increased, labor-saving machines are now part of the landscape. But efficiencies can only reach a certain level before there's a decline of the return on investment. Have we begun to max out the benefits of technology?

I'm an optimist who has faith in a new creative human spirit that will foster renewed hope on the farm. I believe in the art of farming. Good farmers will balance the demands of economics and productivity with the forces of nature. We will respond to weather, not try to control it. Farming in the future means living a mystery; I accept that I won't (and can't) farm the same way every year.

Recently, farm timelines have changed for me. Our daughter, Nikiko, returns home. She's part outsider, not only a woman in an industry dominated by males, but also queer and mixed-race in a world that is too often portrayed as straight, white, and Christian. I watch her struggle with learning curves and witness her response to new challenges. I also bite my tongue, knowing my way of doing things may not fit this new age of agriculture.

She's better equipped to handle change. She doesn't expect to—nor want to—do it all my way. She's young. She's naive. She's full of enthusiasm. All that's exactly what's required for future springs.

But perhaps this isn't much different than when I came back to farm after running away for college. Or when my father took a huge risk following the tragic uprooting and evacuation of Japanese Americans from the West Coast during World War II. He returned to the valley, gambled and bought a farm, and planted family roots.

Is this any more dramatic than my grandparents, who left Japan and sailed across an ocean to farm in a new, very foreign land? They struggled but stayed, their story one shared by many whose ancestors came to California from other places in hope of a better life.

Spring does this to me: I think a lot about what is and what is to be. At the same time, plowing the earth is an ancient rite, a renewal of the past, a ritual others have done for centuries and hopefully will do for many more. Like many, I'm reborn every spring.

NIKIKO'S FIELD NOTE: SHOVEL EDUCATION

The summer I returned from graduate school, I found one of my jiichan's old shovels. The handle had broken loose from the base, and the blade had been worn into a crescent-moon shape from use over many years, but it was still sharp. I set out to fix the handle and give purpose to this tool once again.

I asked for my dad's help; he taught me to use the drill press, to find the right bolt, and voila, in just minutes that handle was firmly reattached. The last step remained: sanding the handle's splintery surface.

I sat on the tailgate of my dad's pickup, sandpaper poised for action. I started swishing the paper back and forth, but I had, without thinking, started at the base of the shovel, where the handle meets the blade.

Dad watched me for a few minutes and then laughed out loud.

"Don't you want to sand the other end? Where your hands will actually touch the shovel?" he asked, a wide grin spread across his face.

"Oh yeah. Whoops." I was embarrassed. How could I have not thought of that?

After a pause, he added, "I guess you didn't learn that in grad school."

Yes. He was right. In returning home, I had entered a new kind of pedagogy. I'd soon realize fully that the model of education I had thrived in—with standards, tests, grades, and clear relationships between teacher and student—were not the ways of the farm. These elements would not help me in this journey. Instead, my jiichan's shovel ushered me into a new space of learning, one that was built on experience, body knowledge, accidental discoveries, flat-out mistakes, and perpetual adjustment to things outside of our control (old equipment breaking, unpredictable market prices, and, most importantly, weather). I am still figuring out this new mode of learning, and I have to remind myself that confusion is fertile. Feeling unprepared might be a permanent fixture of my new existence, and I must grow to embrace that.

GETTING STUCK

Nikiko was stuck. The tractor she was driving had rolled into a mud sinkhole.

It wasn't her fault. Blame it on a surprisingly heavy rainstorm and a sinister place where water always collects. She was working hard, shredding peach tree brush with a big Ford tractor that in one moment was doing just fine and then the next was sunk in the mud. It suddenly looked little, the left tires half buried in the goop, the belly of the tractor rubbing the earth. Machine and driver trapped.

There's something magical about getting stuck in the mud. You're exposed; there's no escaping the problem. On a farm, everyone eventually gets stuck; it's part of the work you do. Anyone who hasn't gotten stuck probably hasn't worked that hard. Overly cautious folks may avoid problems by waiting for ideal conditions that rarely happen. Note: They also don't get much work done. Or they won't work unless they have bigger, stronger equipment—which most of us don't readily have at our disposal. Others never take chances and avoid wet-dangerous-risky maneuvers and hence miss numerous opportunities in life.

Nikiko walked up to me sheepishly, but not too upset, no longer embarrassed. (This wasn't her first time.) She simply announced, "Dad, I got stuck."

I've learned not to overreact. I've been there—both as the "stuckee" and the "un-stucker." My response: "Let's see if we can pull you out."

The act of pulling someone out after they get stuck can be a moment of pure joy for a parent: you get your child out of a jam and still feel needed. Yet in our current complex world, we often lack opportunities to help pull someone out, especially our own children, especially when they're grown. With technology, it's often the reverse—children helping parents after they get stuck. A computer freezes, you can't run the printer, you don't understand Facebook or Twitter. You feel dumb, old and obsolete. I miss the simpler ways the previous generations got stuck in mud. I imagine my father had to confess to his father after getting the wagon or plow stuck and the team of horses or mules simply stopped when they could not dislodge the implement.

As Nikiko and I walked to the suspended tractor, it looked like a trapped animal, exhausted from struggle, lying motionless, awaiting its fate. "You were smart to give up," I reassured her.

"I didn't give up," she responded. Nikiko understood when to accept mistakes. The mature solution was to ask for help, despite her typical drive for perfection.

On the farm, all work is public: you must openly admit you're stuck because it often can't be hidden. When I was a teenager, why did it seem I often got stuck in the row next to the road? Then everyone driving by could witness that Masumoto kid always screwing up. This was the type of rural openness none of us farm kids could avoid. All we could do was forge forward with great determination. (I remember after one rainy winter, a neighbor's kid got their tractor stuck and it sat in the field for a week. Even the local weekly newspaper came out and took a picture. People then drove out to take a look at it. The farmer should have charged for people to stop and gawk at the sight.)

Nikiko walked back to the shed and retrieved another tractor, a newer Kubota with four-wheel drive. I pulled a heavy chain from our barn and dragged it to the scene, smiling as I thought of the numerous times this chain had rescued me. Getting stuck in the fields feels very human. Your work connects you directly with nature—mano y mano—and nature will always find ways to trick and trap you. On the farm, struggle is welcomed; it's often necessary as we grow and grow things.

The world Nikiko is entering will be different and difficult, with new challenges and pitfalls. Already, modern-day fixes are beyond my capacities. Cars are much too complicated to fix yourself when they break; you now need a computer just to tell you what's wrong with the engine. There once was a time you simply listened to an engine in order to diagnose the problem. The ability to mimic the sound of trouble was the best method of communication to a mechanic. The difference between a *clunk, clunk, clunk* and a high-pitched whine and a weak *rum, rum, rum* helped make repairs possible and efficient. Those days have long passed us. Even home repairs are trickier now. Work under the kitchen sink is daunting. Garbage disposals and water lines to the refrigerator and dishwasher get in the way. Plastic pipes are not as forgiving as metal and don't pair well with my cheap pipe wrenches and clumsy hands.

Today, getting stuck in life is also more complex, and it sometimes leads to legal issues and financial problems. Solutions are often complicated and expensive rather than simple and cheap, and you sometimes need the help of a professional, not just "good old Dad."

On a farm, intent matters. When younger, you will get stuck and will eventually have to approach an authority figure, like a parent, with an admission of guilt and responsibility. Later, though, you may have the chance to repay your debt. I fondly recall the first time I had to pull Dad's tractor out of the fields after he got stuck. The tandem disk had snapped a lift-arm pin, causing it to crash into the dirt and dig itself into a mass of weeds and mud. Dad then asked me, "Can you lend a hand?" and I jumped at the chance. And now as a parent, I revel in the feeling that my child still believes I can fix things.

I have to admit, I knew well the location where my daughter got stuck. Hardpan sits right below that spot, and water doesn't drain very well. But more significantly, the ground is compacted from all the times I got stuck there and my father had to pull me out. Perhaps I should consider this spot sacred, a place where rites of passage unfold. A sense of history lingers there. When I got stuck years ago, I was angry. My immature attitude meant I refused to ask for help and tried brute force to dislodge the tires. Of course I failed; stubbornness does not often work against nature. In the end, nature laughed and my father smiled. He attached a heavy chain and pulled me out with another tractor. I calmed down as we slipped out of the mud and passed by the evidence of my struggle: a broken shovel handle, chunks of hardpan rock I had tried to jam underneath the spinning tire, the articles of clothing I had slowly removed and scattered as I'd gotten hotter and hotter trying to free myself. I'm pleased to see my children are not like me. It took a while for me to divulge this story to Nikiko. She seemed to enjoy hearing of my errors.

I now welcome future moments of being stuck with my children. It's an instant that creates a special bond between a parent and a child as we share both the act of getting stuck and the wonderful anticipation and joy of being unstuck.

THINKING SLOW

I keep a journal in which I record notes, observations, emotions, dreams. These field notes work as a management tool, documenting details of cultural practices and business plans. But my recording quickly slips into reflection. My mind isn't a computer, and as I age my RAM gets too cluttered to keep it all in my head. I need a place to store notes to myself. These observations provide the raw materials for a writer. The genesis of a story begins in my notes. Lessons I'll bestow to Nikiko are born in my journal.

All good farmers make keen observations. Some rely on memory; a few, like myself, pretend to work like trained anthropologists. Field notes go beyond detailing work. We can probe and push ourselves with hard questions. The art of seeing embodies seeking and uncovering the truth. I have learned to trust my notes.

I have notes about irrigation, where the highest ground lies, and "the silly lesson that water does not flow uphill." I wrote that after a long, long summer day decades ago. I had labored, cajoled, begged, and pleaded with the water, trying to get some of the life-giving liquid to the thirsty vines. I had not made my furrows deep enough, and only a trickle of water reached the vine row ends. I can reread my journal pages from that tough workday and feel my body caked with dried mud and sweat mixed with some blood.

———————————

"The best decisions are made in the heat of battle." I wrote that while we were harvesting in the mid-1980s. Bad prices. Small fruit. A hundred-degree heat wave. It was decision time—time to admit that things had changed and an old variety of peaches would have to go. I grew weary of fighting those three hundred peach trees that had a history of growing mediocre fruit.

Marcy counseled me not to make a rash decision; perhaps I should wait until the fall before calling the bulldozer to pull out any trees. But that would be too late. By then I'd have calmed down, forgotten the feel in my stomach as I watched fruit that would never develop. They were supposed to grow fat, at least for the modern marketplace demands, but instead they taunted me by staying tiny. "To hell with them," I wrote. I had to be true to myself.

A short time later, I recorded, "I hear the crack of a thick tree limb as it gives in to the roar of a diesel Caterpillar engine. The driver is disguised, thick goggles coated with dust. Man and machine as one, pushing over tree after tree. To hell with them."

Bitter? You bet. A note to bequeath to Nikiko? Yes, and more: a story of killing. Euthanasia as a management and feel-good tool. Those trees were killing the farmer.

———————————

"You become a farmer by dying a little every day." What did that mean in 1997? My father had had his stroke that year. I was forty years old and had been back on the farm for almost twenty years, but I still felt like a kid. "What am I going to do without you?" I wrote. Perhaps Nikiko won't become a farmer until after I die. A new journal entry about transformation.

———————

Lost in our age of instantaneous communication, old-fashioned reflection still resides, hidden in the wave of social media and digital exchanges. Today, we often respond without thinking. Quick texts, emails, instant messaging, pushing keyboard letters without knowing exactly what we want to say. Where do we train ourselves to think?

I think while I'm in the fields, especially during repetitive and monotonous chores. I daydream, allowing my vivid imagination to wander and run wild. I jot down notes that I later take to my basement office to expand in my journals. Writing forces me to think before I commit a thought to words. I want to capture the raw emotions and the wild from my fields.

So I write slowly with a license to explore. Make mistakes. Be silly. Talk about people. Snarky. Cynical. But mostly authentic.

———————

There are entries I'd typically never share with anyone.

When I was first dating Marcy, then a graduate student at UC Davis, she came to visit the farm. It was late September, dreadfully hot, over 110 degrees, and the raisin crop was baking in the sun. I couldn't get workers into my vineyards soon enough; the harvest was over-drying.

I had asked Marcy not to visit that day. I was exhausted, having spent hours trying to roll the paper trays upon which grapes were rapidly curing into raisins. Workers had picked the green grapes about two weeks earlier and set them on the two-by-three-foot trays next to each vine, where they remained in the intense heat. We had about fifty thousand trays that year, each one holding about three to four pounds of raisins.

I labored as fast as I could, mostly during the cooler time in the evening, although it was still 100 degrees at sunset. As dusk approached, Marcy joined me in the fields, where together we rolled each tray by hand, on our knees. Tray after tray. Row after row. Hour after hour. Eventually by moonlight.

Darkness fell and still we crawled from tray to tray, feeling them with our hands. At the end of one row we silently stood next to each other. We had saved a few more trays. We were exhausted and leaned against one another. We turned and gently kissed. At first a quick peck, then a longer embrace as we exchanged the dust on our lips. I smiled and knew I'd live forever with this partner on this farm.

In my journal I called her by a new nickname: Dusty Lips. Later I could recall the moment and smile at my attempt to write a romance novel—my version of *The Bridges of Madison County*, or an old-fashioned take on *Fifty Shades of Grey Dust*. Silly but fun journaling.

My journals allow me to learn perspective. I can go back weeks, months, and even years to understand events in context over time.

In 1983, the year Marcy and I married, we had one of our worst years financially on the farm. We lost lots of money. In my journal I tried to capture what that meant. Most people had no idea of the ramifications of a bad season. Not a decline in the stock market and a squandering of value. Not being unemployed, but instead working the entire year, "as if you paid others for the right to work." I joked: "Imagine working full-time, and instead of an end-of-the-year bonus you get a bill."

I wrote about the embarrassment, the frustration, the humiliation of failure. A moment of regret: was this any way to begin a marriage? A bad year has repercussions only revealed over time. A journal can capture that. A few months later I wrote about Marcy's professional work in education. Without her work, we would not have benefits. Without her efforts, I'm not sure I could have continued to farm. She became my personal financer; she supported my habit and addiction. I could only trust my work if she trusted me.

So I wrote about how my failures liberated her and our relationship. And when Nikiko was born two years later, my role as a parent had been defined: I stayed home to raise our children. On a farm. In the fields. With their father. I made less money, but I could add a different value to our household.

I search for meaning in my journal writing. Farming can be terribly lonely and isolating, something Nikiko also struggles with. In my journals I'm allowed to grieve. Over time there may be healing, but in the moment, writing can help sort things out and find significance.

When Nikiko was very young, perhaps about eight years old, a spring thunderstorm marched into the valley and the hail stripped the peach trees of their blossoms. Depressed, I wrote about it in my journal, admitting my inability to shelter my children from such storms. Farmers draw no lines between home and work; their family is exposed to the elements; they bare themselves and their emotions.

I wrote about running outside, looking up, and screaming to the clouds. I pled for them to depart, to spare me. I demanded to know what I had done to deserve this punishment. I knew my actions were futile, but I needed to respond physically to this physical insult to my farm.

Nikiko witnessed me. Then, through the eyes of a child, she did what she could. She drew a picture of a farmer wearing a big hat to protect himself and gave it to me. I kept it on my wall for years and wrote about it in my journal: "On our farm, I can't keep secrets. Nature exposes. That's why we're a family farm. A journal of stories and truths."

THE ART OF PRUNING

My father taught me how to prune a peach tree. He began with an old tree with weaker branches and gaps where a limb had died and been sawed off. Why a misshapen specimen? Because my sage father knew I was young and learning how to prune; he wisely didn't want to sacrifice a good tree.

Our pruning shears are designed for trees with big branches. The head has a curved blade so that when you cut into a large branch the wood is drawn inward rather than pushed outward, perfectly aligning with the blade. We can snip and cut with rapid, fluid motions.

The biggest lesson in pruning is that trees like to be pruned. They want to be sliced and diced. They need annual haircuts and long to be free of the rank growth. They seek training from the farmer. Pruning marks yet another annual rite of passage on our farm: Cut out the things that don't belong. Purge the negatives. Open up leaf canopies to life. My job is to learn what belongs and what does not.

So I whacked, snipped, and slashed. Branches fell, wood dropped. I copied the motions of my father pruning the tree next to mine. As his tree took form, he looked like a sculptor. Hidden inside the mass of branches and limbs quietly rested a clean, simple tree.

Dad warned me, "Don't be afraid to cut."

I thought you were supposed to take your time, reflect, admire your craft.

"The biggest problem is that people don't prune enough," he said. Then he repeated, "You need to cut out things that don't belong."

But what belongs? I wanted him to explain it in detail, in words, but instead he kept working around his tree, dancing with his shears. His arms at first seemed to flail and swing wildly, and I mistook his speed for uncontrolled and imprecise movements. He glided and shuffled with expert footwork. I noticed he seemed to be looking forward to the next branch while still slicing another.

Even the way he placed his tripod ladder was strategic. These traditional fruit-farming ladders may look unstable, with only a single third leg pushed outward and away from the steps, but they're perfectly designed to fit between the major limbs. With a properly positioned ladder, you don't have to nibble from the outside edges, you can insert yourself into the canopy and start whacking. In order to purge, you must get close up and in the middle. You can't cleanse from a distance; you have to get your hands dirty.

Our trees already have the basic contours, the biggest and oldest branches molded into a vase, not too wide at the top but also not too narrow and tight. The finest trees resemble the shape of a wine glass, gently opening from the stem base and flowing outward. Too wide and the tree resembles a martini glass; it will expose the young wood to too much sun. Too narrow and it looks like a champagne glass, with not enough light for the growing fruit. I concluded that it pays to know your alcoholic beverages when you engage your pruning shears. He grinned and shook his head at my observations.

Gradually, a new tree took shape. Six or seven major limbs grew upward from the base, and each limb had numerous branches with smaller "hangers" extending outward. This skinnier and smaller one-year-old wood would have fruit the next year. (Peaches bear only on wood from the prior year, not the hardened, brown branches from years before.) The fresh hangers are often reddish in color, and you can see the little nodes on the surface where leaves and blossoms await the warmth of spring. Once you identify them, you can't miss them. For years I had never paid attention; I had never paused long enough to see the next year.

But the real magic happened when my father finished his tree. Suddenly I looked up and could see through the branches. I saw the blue-gray sky of winter. It was as if he had opened up a secret world. Trapped behind the clutter of growth, hidden from our view, he had pruned away branches to open up the canopy.

"It's all about light," he explained. He slowly motioned with his hand, raising it up high and then swinging it downward, wiggling his fingers as he mimicked the sunlight entering the treetop and striking the wood where buds lie, awaiting the warmth that signals the change of season. The fruit that gets the most sunlight will taste the best, he told me. "Peaches worship the sun, and so should you."

He explained that you want to imagine the sunlight months from now, you want to *feel summer* when you prune; in the cold of winter, you want to see the light of summer as it penetrates and gives life. His greatest skill in life was his ability to anticipate. Destiny, he seemed to tell me, was determined by your prior actions. Fate was built on everyday, seemingly simple deeds. Tend to your work. Trust your acts. Believe in yourself.

As my father whipped around another tree, opening it up to light, I could see the negative space he had left behind. I no longer saw just the branches but also the space between the branches.

He helped me finish my tree and my lesson. I had more questions, but he spoke little. He was about showing, not telling.

Now it is my turn, and I still struggle every year. It's hard to purge things in our lives. Prune away the excess. We live in a world of accumulation. And I'm not sure I have the vision of my father as I try to see the future in the present.

NIKIKO'S FIELD NOTE: SEEING THE SKY

A few years ago my dad gave me the task of pruning a peach tree. It was the first time I'd ever had to do it alone.

Pruning seems deceptively simple. Armed with loppers and a ladder, I was to do one of three things to each small branch, the "hangers" that will bear fruit the following year:

Option 1: Leave the hanger alone.

Option 2: Cut the hanger back (maybe by as much as half, depending on its original length).

Option 3: Chop off the hanger altogether.

It's just three simple options, but arriving at the right one requires me to consider a lot of other variables: placement of the hanger, the direction it's growing, the overall strength of the tree, the variety of peach, the placement of the tree in the orchard, and predictions on how many fruits we want per tree, just to name a few.

I started snipping timidly, and with each cut I grew more tense. I kept thinking about what was lost with each slice of my shear; I imagined peaches disappearing and I felt the pressure of the job. There are no do-overs.

The next day I watched my dad prune. He worked with ease and comfort, as if pruning away the extra hangers was allowing both him and tree to breathe more freely. He said to me, "What I look for is not the individual branch but the sky."

Observing his artistry, I began to shift my thoughts, retraining myself to focus not on what is lost when we prune but what remains—brushstrokes of branches upon a buoyant blue sky.

It was yet another life lesson I collected from Dad and farming: What is allowed to grow in our lives when we pare down? What are we missing when we look at what is lost instead of what remains? What is the bigger picture?

OLD SHOVELS, OLD FRIENDS

I say goodbye to a dear friend in the late autumn: I put away my old farm shovel for the winter.

I wipe off the dirt and dust and pause to think of what we went through this past year. A nice cold winter. Heavy late-winter and spring rains that resulted in weeds. Lots of weeds; they grew where they normally don't. A wild downpour in early June and frantic shoveling to channel water away from equipment sitting in soft dirt. A summer peach and nectarine harvest with good enough prices. Anticipation of a great raisin year until October rains dumped an inch of water in seven hours. Weeds grew between the raisin trays, and grapes desperately tried to dry.

My shovel and I work hard; it's what we do. As the years pass, I find myself leaning against my shovel more and more. Not just a place to park my hands and arms but to stop and rest. My shovel supports me.

The blade is browned by rust, but I am struck by the fine old handle. The original one snapped long ago, a victim of age and an impatient young farmer. But instead of replacing it, we simply shortened the wood, and now it fits the Masumotos well. Like my father, I'm five feet, six inches, with broad shoulders and small hands, a perfect match with our shovel and the narrow handle that's not too long. My body is built in a way that Nikiko can relate to—less manly and more adapted to the real work on this farm.

An old Japanese neighbor once told me short people make good farmers because they're closer to the earth.

My baachan used this shovel. Her tiny four-and-a-half-foot frame paired well with its abbreviated size, and in her old age she religiously reached for this tool before the others. She seemed to enjoy digging weeds, and I would watch her trudge into the fields early in the morning and spend hours reclaiming the land from the wild grasses. Late in the day she'd wander home, her back bent, sore from the hours of labor. Her arms bulged with muscles cut into her spry body. Hunched over, shovel in hand, she came home from another day of life.

When I say I "put away" a shovel for the winter, I simply store it in the barn, lean it against the wall near the door. No good farmer can be without a shovel for very long. We're compulsive: we see a clump of weeds or rainwater puddling and we reach for the shovel like a trusty companion. My dad used to always carry a shovel with him, resting it on his shoulder with the blade end trailing behind, a profile of a worker ready for work, almost like a soldier.

But winter is a season for pause, and perhaps by moving the shovel from the pickup bed to the shed, I'm declaring a timeout. We both need to rest and recuperate.

It is also a time for repair. I use sandpaper to smooth out a gouge in the handle, recalling the time I flung the tool into the pickup bed out of anger (bad peach prices, uncooperative irrigation water due to gophers, and a grumpy farmer). Or I grind out a nick in the blade, smiling at the memory of chasing a squirrel into my junkyard and believing for a moment that I could hurl the shovel like a warrior's spear and stab the prey. (Instead my fantasy ended with a clunk and the sound of the blade bouncing off a stack of old metal pipes.) This brings to mind another time, when

I was amazed the shovel didn't break after it fell off the tractor (I was driving too fast) and I ran over handle; I was blessed with soft, fluffy, sandy loam soil. No friend should take such abuse. Next year I vow to be wiser and more careful.

I pause to admire the sheen of the handle. I stroke the tight grain and feel a coolness in the sleek surface. Sweat and body oil from working hands has polished the wood tens of thousands of times.

The scent of my late father is in this wood. Years of dedication. Not a monument to his life, not a permanent marker to celebrate his presence. His body has become part of my shovel.

Only when I use the shovel do I truly understand its value. When I poke at weeds, the brown steel slides easily into the dirt, cutting just below the surface, slicing out the roots in a single smooth pass. The rust is deceiving; the metal is still sharp with a filed edge.

But it's the shape of the blade that makes the real difference. The shovel face—a relatively flat piece of steel gently bowed upward at the sides—resembles two crescent-shaped curves that glide through the moist ground of spring and the damp earth of freshly irrigated summer furrows. The blade can swim just below the surface, slicing through delicate roots. Nature has sculpted the correct slopes and angles from solid metal. Generations have honed the proper form, each pass in the sand and silt like a natural whetstone. I can't tell if the original shovel had this shape or if, as years of use have slowly ground the metal to half the size of a new shovel, only two rounded cheeks remain where there might once have been a point.

When I work with this shovel, I work with a piece of my family's past, a gift I inherit. It's a timeline of our family on this land, and I can measure years by the gradual abrading of the steel.

My grandparents and parents all left their marks, the shovel shortened by an inch or more with each generation. When Nikiko inherits this shovel, how will she leave her mark? The legacy of a work ethic, a tool is bound to the land it works.

I pause and think of my contribution. Another half an inch? Actually, probably only a quarter of an inch at most: I don't work that hard. The shovel doesn't lie.

IF STEINBECK WERE A FARMER

Over seventy-five years ago, John Steinbeck published *The Grapes of Wrath,* a tale about the plight of a displaced family's journey from the Dust Bowl in Oklahoma to the Central Valley of California, where they hoped to find work and a new life. When I first read *The Grapes of Wrath* in high school, I was stunned to see a side of our valley, our farms, and our people that I had not imagined—our land of plenty contrasted with the struggle of these immigrants, their constant challenge of poverty, of trying to scratch out a living in our fertile dirt. Hunger in a land where growing food is driven by business and money.

Steinbeck wrote his seminal work during the Great Depression, an era of profound suffering. *The Grapes of Wrath* examined the economic forces that threatened to crush the human spirit and its search for dignity. But what if Steinbeck were a farmer writing today? How might we re-envision his words for our modern circumstances?

If Steinbeck were my neighbor, we'd trade stories, trying to make sense of this crazy world of farming. We would talk about an economic system that drives the price of food lower and lower while farmers struggle to make a living. When *The Grapes of Wrath* was published, in 1939, Americans spent 35 percent of their income on food, and today it's under 9 percent. There is a seemingly never-ending demand to grow food cheaper and cheaper, but what's good for the consumer squeezes us farmers.

We need to work harder every year just to break even.

Nikiko's conversation with Steinbeck would be even more political. They'd share stories of social justice and the struggles of the working class. From their seats on our farmhouse porch, their statements would grow more and more radical as they debated the rights of workers, the mood more and more contentious when deliberating the responsibility of landowners. The slogan "Onward" might prevail in their fight for the disadvantaged. I like to think that, after hours of debate, a feeling of harmony might settle in as the stars greeted them, the peace of the moonlit fields capturing the reflective nature of two passionate thinkers.

Character would be a common topic for all of our farm talks. Readers often forget that the Joad family in *The Grapes of Wrath* were displaced farmers and were not always farmworkers. They arrived in California seeking a place of their own, a story repeated over and over in our valley. My grandparents arrived in the early 1900s from Japan, aliens in a foreign land, and they, like the Joads, had been displaced from their family farms, economic refugees with few other choices. Once in America, they faced a hostile land rife with discrimination and governed by the racist Alien Land Laws that prevented, specifically, "Orientals" from land ownership.

Steinbeck the modern farmer might today write a tale of newer immigrants—the Hmong and other refugees from Southeast Asia, or the Sikh farmers from Punjab, India, or the Armenians from eastern Europe. Instead of documenting the clash between farmworkers and farmers, he might pen a new novel about class struggle: the small family farm verses the machinery of capitalism. The destructive power of economic privilege over the "little guy."

Steinbeck the farmer might write about the crushed hopes of those working hard to achieve a fleeting dream of prosperity. I once wrote about finding torn lottery tickets in my fields, evidence of discarded hopes from our workers. I don't play the lottery—the odds are stacked against you. But I also can't share the same dreams as my workers. As they remind me, I already own a farm.

If Steinbeck were a farmer, race would also be a major issue in his pieces. When *The Grapes of Wrath* was written, America in the 1930s was 90 percent white and only 1.5 percent Latino. By 2010, whites were 72 percent and Latinos had grown to 17 percent nationwide, and in California, the number of whites and Latinos was almost equal, with Asians accounting for 15 percent of the population.

Over the past century, waves of immigrants have filled our valley's fields, and cheap labor has fueled the rapid and successful expansion of our agrarian base. This creates tension between those at the bottom—usually immigrants—fighting for the scraps left for them.

Yet all these characters may have journeyed a common path: the search for home. An eternal optimism permeates our lands; we share a collective narrative that tells of the human spirit's determination to work the earth and plant roots. Dreams are embedded in the hardpan of our soils.

Steinbeck's twenty-first-century Joads would likely be a family from Mexico who had crossed both physical and emotional borders to seek a new life in California. Their journey would be along the back roads, perhaps even Route 66, which led the original Joads to the West Coast. The new Joads would become part of an invisible population whose forgotten hands continue to pick and harvest our foods. Steinbeck the farmer would be fighting for immigration reform, knowing firsthand the role farmworkers play in our food system.

Nikiko would question Steinbeck about his development of other characters. She would demand that women play a more major role. (As I see it, Steinbeck's world often shows women carrying their own weight and working side by side with men, and I'd contend that Ma Joad represents the complexity of being a mother and head of the household.) But there are no queer or gay characters in *The Grapes of Wrath,* and to exclude them would be a lost opportunity in writing today's new farming story. Nikiko might demand a liberated woman's voice be heard and never forgotten. The character she imagines would faces challenges, ride home in triumph, and call out for others to join. While the original *Grapes of Wrath* pushes our thinking about social justice and income inequality, Nikiko would insist environmental issues also be folded into the narrative. You can't separate these elements; they're part of the new world of farming from the perspective of today's generation. John Steinbeck, welcome to the age of the millennials.

I know for sure that Steinbeck the farmer would still be passionate about writing stories of connection, about people who are not rich with money but wealthy in relationships. He would share stories of people on the land, and our quest to scratch out a living in this green valley of dreams. We'd end our conversations with a handshake and a silent nod of the head, acknowledging that much has been accomplished, that more tests and quests wait beyond the horizon, and that ahead lies another day of work, another day of sweat and struggle and hope.

CHANGE OF SEASON

Spring stirs early in California. In March, peach blossoms blanket our farm. Sunlight shines through the translucent petals, creating a pink hue against the brilliant blue sky. The rough, gnarled bark of our trees contrasts the delicate flowers, the old and new side by side like generations sharing the land.

I'm also stirred by surviving another winter. I remember my father waking early each winter day on cold, damp, foggy mornings and, after breakfast, putting on a jacket and gloves and heading out to the vineyard to prune. He was older, in his late seventies, and had suffered a stroke but survived to work his fields. I watched him and realized he was trying to fool nature. By working each winter day, he was making a pact with nature, pruning as if to announce to the world he would see another spring and the life arise from those dormant vine canes. As I grow older, I understand this more.

I wake before sunrise, anxious to make my morning rounds. My ritual begins with a cup of coffee and a quick study of my farm journal, reviewing the work to do and the rhythms of seasons past. We farm within the shadow of those who worked this land before us. Trees and vines have been shaped by many hands performing the annual ritual of pruning. From the porch I can see the orchard my dad and I planted forty-five years ago and the red barn where horses once slept during another era. Another farmer planted the hundred-year-old vines in front of our house, a denser block with vines only six feet apart instead of the typical seven feet. Everything affects how we farm today and how we plan for the future.

I head out into the fields, anticipation lingering in the air. Nikiko joins me, driving over from her house (the same home I grew up in) a quarter mile away. A dog runs alongside her vehicle, the one of our four dogs who had opted for a sleepover at her place. The creatures all greet each other at the crossroads, the excitement of a new day flowing in their yelps and leaps. They know how to celebrate another spring morning as if it's their last.

As she nears and then her car rolls to a stop, I pause to appreciate the moment. A child returns to the land, a young woman steps into a male-dominated world, a father and daughter explore a new relationship, a millennial brings new energy with an attitude. Nikiko believes our work must be transparent and our story told through social media: we grow public food for people, not merely a commodity for consumption.

While I admire her drive to be open, part of me feels uncomfortable. I value the privacy of working alone. I fear there may be no compromise and we will farm with hard questions and with hard heads.

We survey our land by walking and driving in my old farm pickup, assessing the bloom and the work to be done. Farming is physical; it breaks your body and drains your spirit—always more work to do and challenges to face. The economic reality of long hours and low pay lingers on the horizon. With age, I've learned about constraints; I anxiously watch Nikiko dream without limits.

We return to the farmhouse kitchen for a quick, simple morning meal. Marcy prepares for her city job as an educator, and

Korio, our college-age son, drags out of bed for class. We gather with our peach jam on scones and Marcy's canned peaches. (I eat them in a small bowl, others mix them in oatmeal.) We start our day with preserves, a real taste from the farm. I love savoring the summer harvest in winter or early spring. The bounty shared in the off-season by the magic of food preservation is an almost lost art.

Marcy's Wisconsin family knew well such traditions. Our pantry is full of peaches, canned and pickled, transformed into jams and chutneys. In the freezer are frozen sliced peaches and purees that will be reborn as winter peach cobbler and adult mixed drinks. We can celebrate our farm's abundance throughout the year.

I remember that my own parents began every morning with peaches my mother had frozen from the previous summer. Every morning. The treasures were some of our sweetest, often from slightly bruised and overripe fruit we could not send to market. Initially we lamented that these gems couldn't be sold, but then we realized *we* could enjoy them throughout the year. Why not serve the best to ourselves?

During a simple breakfast, I pause to enjoy the fruits of our labor and to reflect about things worth saving, trusting our work has meaning. As the flavor of peach dances on my tongue, I allow myself to believe in the value of my calling, and now *our* calling.

The spell is quickly broken, though. Eighty acres of certified organic peaches, nectarines, and grapes for raisins. On this farm, we don't initiate the day lounging under the canopy of peach blossoms, seated at white-linen-covered tables, sipping peach Bellinis out of champagne glasses, chatting about high art, high stock markets, and hired help.

We're off and running, Marcy to Fresno State University, where her work provides financial stability for the farm. (If it were summer, she'd first pack peaches in our shed, then later in the day jump into the world of higher education.) Korio enjoys college life but can't escape a list of farm chores awaiting him. The farm is never separated from our daily routines.

Nikiko and I plan our workday. In March we prepare to thin the stone fruit. Within a few weeks, the blossoms will transform into tiny fruit, and we will then proceed to thin, or "knock off," 50 to 80 percent of the crop in order to produce larger and better-quality fruit (otherwise we would have golf-ball-sized peaches, more pit than anything else). It's all part of a cycle of anticipation: when you grow things, you always have to be forward-thinking.

My father told me not to look down when we thin the fruit, not to focus on the thousands of tiny green peaches discarded on the ground. They look like corpses following a tragic weather disaster. Instead, he said, "You need to keep your head up, look to the sky, and trust the future." I believe that was his philosophy for raising children, a gentle optimism that we'd all turn out just fine no matter what happened.

He never asked me to take over the farm. Being the youngest of three, no one expected I'd return. I ran off to UC Berkeley and studied sociology and later escaped to Japan for two years. Yet, with that old farmer wisdom, by allowing me to leave, my father provided me with the opportunity to come home. If I'm lucky and half as wise, I can also pass some of this wisdom to my children.

Nikiko and I rope some peach tree limbs that will eventually droop with the weight of ripe fruit. Every tree needs support and shaping. I remember watching Japanese gardeners gently pulling branches to shape a bonsai tree, attaching straps to coerce nature. It took years to properly sculpt a tree, ropes and lines anchoring a major scaffold, the cord stretching to the ground, secured by mooring buried in the earth. I asked the old men if they tried to hide the cords to keep the distractions out of sight. They smiled and responded, "What cords?"

I use work gloves, part of a farmer's toolbox. This particular pair is old, and they once had holes in the worn fingertips, but I cut off the tips for roping and tying knots, to more easily manipulate the strands with bare fingers. After my father passed away, I used his old work gloves for years, but I altered and recycled them as I took over the farm, often clipping off the stiff leather ends to expose my fingertips for intricate work. I could smell the scent of his sweat in the aged hide and was comforted by his presence as I stumbled through my own quest to learn. I was no longer the son, I was now the farmer.

During the workday, I will often snack, sometimes nibbling some of our dried peach slices. Marcy and Nikiko have perfected a method that takes advantage of our 100-degree summer days to cure the thinly sliced peaches and seal flavor into these treats. No processing, no additives, just real and authentic. I like sucking on the wrinkled slices; the flavor stays with me.

I may pause for a midday meal, but often my schedule is based on work demands. I will start to repair broken farm equipment and notoriously underestimate the time required.

I remember that when I lived in Japan and worked on the family rice farm, they'd bring out bento boxes to eat in the fields. The lovely packaging and efficient containers held the promise of an excellent rest from work, an escape from the sweat. But in America I fear work consumes us and dominates our schedules. I foolishly skip meals, believing tractor maintenance is more important than my own.

Yet farm equipment must be repaired, and breaking things is part of how I grew up on the farm. I had the perfect relationship with my father: I'd break things and he'd fix them. He knew how to weld and repair, how redesign a disk or plow like a sage engineer, accommodating my youthful impatience, such as when I drove a tractor too fast or accidentally bent implements.

Nikiko, too, will learn these life lessons—when to slow down in the field with hardpan rocks, how to raise the depth of the disk in the heavier clay pockets of our land. She will break things and accept being dependent on someone else who will fix her mistakes. And ultimately she will learn how to fix things without a father.

While I work, I always make notes. This is how I write—jotting down observations and ideas, thoughts and feelings, emotions and reflections, all on little 3x5-inch cards that fit perfectly in my shirt pocket. Some notes are straightforward and simple ("Warm spike in March temperatures" or "Sun Crest peach tree, row 2 west of center—needs grafting"), while others are contemplative. I once wrote, "Keeping track of the age of tractors not by the miles they were driven but by the hours they were used." Tractors don't have mileage odometers; they use a meter that records the number of hours used. It's a different way to measure age, less quantitative and more qualitative. My note continues: "How different would our lives be if we measured experiences by hours?" Imagine the time dedicated to friendships or parenting. Periods allocated to playing. Hours consumed by worrying. Moments of depression or joy. A pause for humility.

I think in stories, and numbers alone can't capture the significance of a family farm; what is valuable here has to be grounded in narrative. We're all characters in an unfolding novel—

sometimes a mystery with a few tragic moments—and the brutal reality is that often people pay little for what feeds their bodies. Stories fit how we work this land. We believe food has meaning beyond prices, nutrition labels, and profit margins.

How will Nikiko measure time on the farm? I expect she'll create her own scale, her own priority agenda, and it should be much different than mine. The challenge will be accepting her wishes and only offering my opinions and thoughts. Or not. I learned the most by observation and repetition, a practice of study by osmosis. The lessons I mastered were the ones I personalized. I progressed furthest in my learning when I was able to redefine a task or process and make it my own. I hope I can let Nikiko do the same.

Perhaps that's why we work both side by side and separately in our own fields. Farming in general has become less communal. Mechanization and new technologies are often designed for an individual rather than a team approach. A tractor is driven solo, no farm implement requires two people to operate, and even hand labor usually has each person working at his or her own pace in his or her own row. Yet I do some of my wisest thinking while alone, and cultivating a relationship with nature seems best done one-on-one. The land demands our full attention, free from distractions.

Near the end of my workday, I shovel weeds and my thoughts wander freely. I'm too tired to be calculating, so I seek comfort in little things like an old shovel. Tonight my body will ache; every March, spring training becomes more challenging.

How many harvests do I have left? Even as I watch another generation on this land, I still feel young. Sore knees and strained muscles tell me something different, however, and I struggle to strike a balance between growing food full of life and working so hard that I miss out on my own life.

Still, I thrive in this world of real work and shun those from the world of privilege. On our farm, we remain part of a working class, engaging in livelihoods where we are judged by our deeds, not by our bank accounts. Values matter here in a way they don't elsewhere, and as I age I find myself in a crusade to champion the middle class and those less fortunate, while challenging the elite. As a day ends, I trudge home weary yet full of hope.

When I return to the house, Marcy and Nikiko will often be preparing a simple feast for dinner. Food traditions embody our meals—Marcy drawing on her Wisconsin family recipes, such as pickled peaches, and Nikiko investigating the boundaries of tradition and innovation with panko-fried peaches. Marcy has learned a secret—some of our peach varieties explode with character by simply adding heat. Nikiko explores inventive pairings, like peaches and pork. I help with my own talents: I like to do dishes.

Tonight, we have invited friends and family to join us. We share a salad made with local ingredients and fruit from our backyard and that of a neighbor. Each course is served with stories. Voices fill the room, and around the table our conversations mix with the food memories. This is the proper way to complete a spring farm day: an old-fashioned gathering at a table full of optimism.

NIKIKO'S FIELD NOTE: PASSING

I was making my routine call home to my parents from my small studio apartment in Austin, Texas, where I had moved for graduate school. My mom asked me how I was doing, I said great, and then my dad came on the phone, his voice with a soft tone. Something had happened to my jiichan. He wasn't doing well. It was unclear what had happened exactly; it had been more than a decade since his first stroke, and later a second stroke had put him in a wheelchair. Now, something else was wrong.

He was at home in hospice care. I wasn't sure what that looked like. Was he hooked up to odd electronic medical devices? Was he conscious? How weak was he and how fast was he declining? Could I do anything to help?

I hastily began to pack a suitcase, then paused as I looked at my closet. My hand ran over a black dress. Deep breath.

As planes rarely do, mine landed early. My mom was in the car to pick me up, and we greeted with our arms wrapped around each other tightly. My mom sighed and offered a muted, "Welcome home."

When I walked into the house, my baachan looked up. Our eyes met, and her face was at first confused, but as soon as she registered who I was, I saw a wave of joy sweep over her. But then, just as quickly, she was overcome with a look of loving sadness. As we embraced, her voice broke as she said, "Oh honey, thank you for coming." I took a deep breath and was ushered by the warmth of my dad, mom, and brother to the hospital bed now in the living room, in front of the family's Buddhist shrine.

He looked smaller than I remembered him. I had lived with him and my baachan for two years before going to graduate school, and I knew his warm smile and his strong grip. There was never a day when he wasn't at the door waving and smiling when I walked up to the house.

Now, he lay very still, his legs swaddled by a blanket. As I approached, my dad gently nudged him awake. He helped open his eyes and then said, "Look who's here."

"Hi, Jiichan." I extended my hand and clasped his, not like a formal handshake but like the greeting you'd see between brothers, between equals, each palm embracing diagonal to the other. The corners of his mouth turned up in a small smile. He chuckled momentarily before a roaring cough claimed his lungs.

For the first time ever, I was grateful my baachan was deaf and couldn't hear her husband struggling to breathe normally. She was preparing lunch for all of us, still caring for others even in the midst of her sorrow. After Jiichan fell back asleep, we all sat down together to eat: my baachan, my brother, me, my mom, and my dad. My Aunt Shirley was headed to the house, just minutes away. The family had been taking shifts.

From the next room came more coughing, and we all looked at each other. My dad sprung up to attend to Jiichan. More coughing. And then I heard my dad call out in a tone I'd never heard him use before. He asked for his partner, my mom, his voice unsure.

I heard some whispering from the next room. The coughing subsided and silence descended.

Dad softly summoned us kids over. "I think Jiichan just passed," he said. I saw Jiichan's mouth releasing small puffs, his last breaths returning to the world he was no longer part of. I saw his last exhale, after which there was no inhale.

Dad asked us to bring Baachan over.

I don't remember the words or gestures used, but we signaled to her that her husband was gone.

She blinked, bent over his body, and looked closely at his face. She stood back up and said definitively, "No, he's sleeping." My dad slowly shook his head.

Her voice grew louder. "No, he's sleeping." She clutched his shoulder. "No," she repeated, this time slower, this time as if in defeat.

She leaned over to his face and kissed him. We stood around his body, holding each other. Who can ever be prepared for a moment like this?

A few days later, the extended family gathered at his viewing. I volunteered to scribe for my baachan so she could interact with everyone. One of my dad's cousins who is a nurse came up and gave us hugs, and then she leaned over to me and said she had heard I'd arrived just twenty minutes before he had passed. "Sometimes they wait for you," she said.

I'll never know. But I don't want to believe my jiichan was waiting for me specifically. I think he was waiting for us all to be together.

When I returned to Austin to finish the semester, I approached my schoolwork with a new sense of urgency. I declined to move forward into the PhD program and instead came home as soon as I finished the master's program. I didn't want to wait any longer to start my farming career. The place, our home, our farm, became a shrine, my work a ritual practice of being close to Jiichan again.

PART THREE

THE FUTURE OF FOOD: WHAT'S NEXT

DO YOU HEAR THE PEOPLE EAT?

I complete a ritual done every five years: the farm census.

But this isn't my father's or grandfather's census. This one appeals to a new audience, and they're not farmers. Foodies and the food community have their fingerprints on this document. This farm census contains a hint of what may come: a new urban and public interest in agriculture. This bodes well for my farm—and Nikiko's future.

I see the face of "big ag" and also "new ag" on the census pages. Beyond the dirt and dust of our farms, city folks are more than ever a part of what agriculture is becoming. We are gradually entering a new political world in which policies like the Farm Bill are no longer relegated to conversations in rural coffee shops or the corporate offices of giant agricultural companies. I have friends outside of farming, but especially those who are "into food," asking me about this seemingly obscure government report. Things have changed; our farm doesn't seem so provincial. A small farm has been transported into the public arena.

In years ending in 2s and 7s (e.g., 2007, 2012, and 2017), the USDA gathers data from farmers about their operations. The theme for 2012, "Your Voice, Your Future, Your Responsibility," urged all farmers to participate. "'Tis the season to be counted," read one press release.

The farm census began 150 years ago with a complete count of every farm and ranch in the United States. It was the only source of valid information for the industry and government programs. For decades, only the ag community paid attention to the data about how many acres were farmed or leased, how much land was cropland versus pasture, how much earth was irrigated or dry farmed. Officials solicited specific information for each type of farm or ranch, including acres harvested, quantity harvested, and value of sales. Animals including cattle, hogs, goats, poultry, and aquaculture were tallied; these days the data also includes bees and Christmas trees. Other sections ask about farm labor and production costs, and the use of fertilizers and chemicals.

By law, I have an obligation to fill out this form. Typically I race through the packet (in 2012 it was twenty-four pages long), since many of the questions are not applicable to our farm. (I don't make maple syrup or grow mushrooms.) But that year the final five pages grabbed my attention. The questions represented a shift away from the standard inquiries about production and volume. I sensed a change in attitude, an indication that farming is no longer just about production but about relationships. I saw recognition that the true nature of farming isn't measured by bushels or tons but by people and their approach to working with nature.

Organic agriculture now has its own section, as does renewable energy. One remarkable question asked about "value-added crops," which include on-farm processing and packaging of products, like jams or preserves, that allow the farmer to capture a larger share of income, as well as farm-related activities such as direct marketing and agritourism. Another query sought information about CSA, or community-supported agriculture. A final section explored direct food sales for human consumption.

Is this the moment to step out of the shadows and at last bask in the light of public recognition?

Annual sales from organic farms account for more than $28 billion (according to the Organic Trade Association's latest estimates) and comprise 4 percent of all food sales. We organic farmers may be small players in the world of industrial agriculture, but our growth rate remains one of the fastest in all sectors of agriculture.

Organic farming data was first gathered in the 2007 census and a special supplemental survey. Now the trends can be examined over time. Before this, no one knew the exact numbers, and most talk was anecdotal and speculative. Finally this small and often ignored sector is developing clarity, and perhaps newfound legitimacy. Each question I am asked about my organic farm feels like a nod of approval.

And that's exactly how I feel: accepted. Someone might analyze my enthusiasm and summarize, "Talk about an inferiority complex!" Perhaps they are right—so much of our farm's work has been overlooked, disregarded, and ignored for decades. But by asking about our farm, the census proclaims that we belong, we matter. We are asked to be counted. It's a small victory that few others may notice, a significant moment that one generation celebrates and another may overlook because organics has become commonplace.

The form mined for data about the "buy local" movement, as reflected in questions about direct sales, including farmers markets, roadside stands, and CSA programs. The 2012 census also specifically asked about value-added products, including jams, preserves, cider, and wine. Even though they account for less than 1 percent of food sales, this seemingly small and insignificant sector contributes to a new farming culture that thrives on direct interface with the public. Perhaps our farm's worth is not measured by how much money it makes but instead by the fact that it is even worth counting.

It should also be noted that in farming, as in other industries, a small business can still have major influence on the market. Consider that Apple controls only 11 percent of the computer market, yet it's viewed as the leader in design and innovation, its products and philosophies rippling throughout the high-tech landscape. Similarly, the organic, buy-local, and direct-sales sectors of farming may be small, yet they lead the charge with innovative ideas that inspire changes in how food is grown, marketed, and consumed. The public has a new stake in our nation's food, and "new ag" is discovering ways to connect with these consumers.

For example, in California, a new wave of small but perhaps significant growth is unfolding with the passage of the Cottage Food Act, which went into effect in 2013. This new category of food production allows smaller operations to work out of home kitchens so long as the products are non–potentially hazardous. This includes pies, preserves, candy, and baked goods. With this law, our state will create a new generation of food entrepreneurs and connect thousands of growers and farmers directly with their food-loving consumers.

Similar shifts have occurred nationwide since the 2007 farm census, and the changes are showing up not just in progressive states but all over the country. Who would have imagined Iowa would have one of the highest numbers of CSA programs in the nation? Or that Kentucky would be among the leaders in value-added commodities. (Could it be linked to a growing craft bourbon industry?) Could this "good food" movement trickle upward to eventually affect the entire agricultural food industry?

As I send in my farm census, I do so knowing that a new audience awaits beyond the traditional number crunchers and policy wonks. The census will count the new face of agriculture, one that responds to the surging public interest in food. Future forms will be filled out by Nikiko, her voice joining with others, and if the present trajectory continues, new questions will help her realize she is not alone in this world of farming, that the schism between rural and urban may be further reduced by the common bond of the food we eat. Suddenly, the big issues we struggle with on the farm—that too often our voices are unheard, our faces neglected, our work devalued—may not be as relevant when everyone cares where their food comes from.

I feel proud to be counted, and hope people are starting to care. Though it sounds impersonal, it's nice to know there will be data about us small, sustainable farmers. Praise by numbers. We are embraced, sort of.

The three million farmers in the United States make up about 1 percent of the population. The "new ag" represented in the farm census makes up only about 1 percent of that 1 percent. We are a subset of a subset. But we believe that our 1 percent can change what a nation eats.

SEXY FARMERS

Can farming be sexy? According to researchers looking at how agriculture spread across Europe centuries ago, the answer is definitely yes.

Farming is one of the most important inventions in the history of humans. With agriculture, people gained greater control over their food supplies and were no longer forced to wander in search of sustenance. As these growing populations settled lands, they created the foundation for civilization as we know it today.

We know the idea of farming spread around the world, but *how* exactly did that happen? Were farming techniques simply shared and adopted because they worked, or was there something else at play—a human element that made the process more personal?

As a farmer, I ask myself how the next generation will learn about farming and, more significantly, what will they need to know to thrive. Are bright ideas simply shared and adopted, or is there something else at work? Does tradition matter? Are the old ways relevant in the farming industry of the future? Elders traditionally played a role in the transfer of wisdom, but is that a valid system today? Will they teach the next generation of farmers, or will the knowledge come from other sources? Does it even matter who does the teaching anymore? Does Nikiko need this old man when she has all-knowing technology at her fingertips?

I'll trust history in this debate: ideas are powerful, but so are relationships.

According to a study by Patricia Balaresque of the Department of Genetics at the University of Leicester in the United Kingdom, farmers spread throughout Europe over the course of centuries, bringing their new ideas with them. Her study focused on DNA and gene pools, beginning with populations in the Near East. The Fertile Crescent of Iraq and Syria has been documented as one of the origin points of farming. Over centuries, agriculture gradually spread from there.

But did farmers themselves or the idea of agriculture migrate?

If the technology of farming spread, and not the farmer, then clusters of native genes should have survived. Genetic patterns should have been erratic. But Balaresque points out that men across Europe have remarkably similar Y chromosomes. There seems to have been a pattern of a smooth march of male farmers' genes across the continent.

The Y chromosome can be classified into different lineages, which reflect a person's geographic ancestry. The most common Y chromosome lineage in Europe, carried by more than one hundred million men, was passed down from father to son with very little change. It follows a gradient from southeast to northwest. This "molecular clock" paints a picture of genetic spread across Europe from Middle Eastern origins, where farming began. More than 80 percent of European Y chromosomes descended from migrating farmers.

This means that farmers migrated along with their ideas, which transformed civilization. Ideas alone did not spread without the good old farmer. I like that theory.

I imagine Europe ten thousand years ago, when agriculture was first embraced. Although this revolution was gradual, the slow adoption of farming techniques allowed larger populations to grow, village structures to evolve, and new economies of trade to flourish. Early farmers journeyed to neighboring regions and intermarried with local women. As Balaresque theorizes, "To us, this suggests a reproductive advantage for farming males over indigenous hunter-gatherer males during the switch from hunting and gathering to farming." She adds, fancifully, "Maybe back then it was just sexier to be a farmer."

In my wild imagination, I picture the earliest farmers as meek-looking individuals, more comfortable with digging sticks than with clubs. They probably enjoyed the social life of farming, frequently returning to the same plots of land, walking along dusty paths, talking with their neighbors and complaining about the weather. They disdained the man caves and longed for the fresh open air. They stayed put and planted roots.

I fantasize that local women were more attracted to these stable farmers, who nurtured things and stayed at home. Perhaps farmers could better remember birthdays and anniversaries, since they were used to utilizing seasonal calendars. They may have also shared their emotions about violent hailstorms or killer frosts and stood side by side with their mates to fight off pests and predators. Maybe these farmers even held hands with their partners, understanding and acknowledging the value of calluses. I hope that they treated their companions with respect and understood the value of partnership and working together in the fields. It could have happened in my version of the past! For once, the nice guy wins.

Certainly, emotions are not part of a DNA study, but I like to think love plays a historic role. Love, the romantic mover of heart and soul, like the flowering of a rose or, in this case, like a basic grain spreading across Europe. An agrarian rhythm of slow dancing—with nature and with mates—suited well to matters of the heart.

Nikiko, of course, would claim it was women who mastered the art of growing things because they have more patience than men and an ability to see the world holistically. Together with their partners—sometimes other women—they could transcend the myopic and hardheaded thinking of men.

Does this history of agricultural tradition have any significance in the world of farming today? How are advances in technology adopted by others, and will that trend continue? The modern agricultural world has changed drastically from the world in which my father grew up and worked. The rapid pace of change—in everything from research to farming practices to marketing—has been startling. To think my family once farmed with mules and horses just a generation ago staggers my reasoning. The current rapid adoption of new techniques is not the slow, methodical process of farmers migrating and spreading their ideas along with their seed. Today, digital technology—computers and the Internet and all that comes with it—have impacted how we farm, catapulting us into unknown territory.

But I believe the human element still matters. As the food industry embraces the information age, how will the work of farmers be linked to the new knowledge economy? Social media frames our foods in very different ways than ever before. The new food revolution represents not only the spread of ideas from farmer to farmer but also the connection between farmers and the public. We have, in this way, not taken humans out of the equation but added more of them to it. Organics, sustainable farming, alternative marketing, farmers markets, community-supported agriculture, and the "buy local" movement—all of these more fully integrate the human factor into our food chain.

And in the future, I do see us "intermarrying" with other non-farm, urban groups who take a keen interest in what we do. Farmers can no longer just grow crops and passively hope they have a buyer for their harvests. We can now transcend the isolation of rural life and thrust ourselves into a new, fertile world of relationships that defy physical boundaries.

I see Nikiko and her millennial generation leading this charge to redefine relationships. She already has—by forcing me to engage with the public more, to grow comfortable with social media, and to search for ways art and our farm will work together to express and engage. Thankfully, our farm is supported by new partners who value what we do and who we are: with their help, others can immediately learn about our story and reach out to us. I know you, you know me. People want to foster a relationship with the people who raise their food. The farmer is a real and authentic character. And, in the end, I hope you like me.

Perhaps it remains true: farmers are still sexy. This makes me smile in my fields.

NIKIKO'S FIELD NOTE: A FARM WOMAN

You are not born a farmer, you do not become one, you are always becoming.

I spent years exploring similar dynamics in college, where I had the privilege of studying feminist theory and performance—not just "performance" as physically embodied on a stage but also in the theories and practices of how we perform identities every day.

In my undergraduate career, I was first introduced to feminist theories of gender, sex, and sexuality through the writings of people such as Simone de Beauvoir ("One is not born, but rather becomes, a woman"), Sojourner Truth (whose incredible speech "Ain't I a Woman?" underscored how race historically—and still today—is enmeshed with gender), and Judith Butler (who wrote about how performances and language police our bodies into gender binaries). I have taken these lessons to the fields as reminders that any identity is not predetermined. While it is true that who we are involves both complex and simple concepts, behaviors, language, ideas, and categories, many of which are constantly taught, reinforced, and rewarded, we are also bound by them. As Michel Foucault would have us remember, where there is power, there is also room for resistance and change.

If none of that made sense to you, that's okay. Keep reading.

These ideas about gender, sex, sexuality, and race are wrapped up in who we think of as farmers. Even the "we" of that sentence matters and is worth defining; I am specifically challenging us in the United States, where we think of farmers as straight white men.

In 2013 I saw a commercial for a Dodge pickup that showed images of people in various rural settings—mostly white people, mostly men. The voiceover was a recording from radio commentator Paul Harvey's piece called "God Made a Farmer." Even without the visuals, the words gave a clear picture of Harvey's idealized vision of a farmer: he is decidedly male and Christian, implicitly white and implicitly straight.

I am none of these things. And I am a farmer.

Feminism asks us to challenge concepts because concepts matter. Concepts—and the language and images that represent them—guide our dreams and aspirations. They build our individual futures as well as our collective one. If we are only shown it's "natural" or "normal" for certain people to do certain things, to choose certain professions, we are constructing a conceptual barrier that makes it more difficult for things to change, for excluded groups to be included.

It is from this perspective that I believe feminism is essential to farming. We know that the nation's population of farmers (still mostly straight white males) is aging, and as we build the next generation, we must recognize that now is the time to dismantle the concepts and structures that make it more difficult for certain types of people to fill their shoes. This includes immigrants, women, people of color, queer people, and other marginalized groups.

I don't want "farmer" to be a stable category. It should change as we change. I am continually becoming a farmer in part because I still have to negotiate others' perceptions about what that means. (I once introduced myself as a farmer to someone who responded with, "Oh, you must be on the finance side," because his limited and sexist imagination could not conceive that a woman would/could/should do anything outside an office.) And I am continually becoming a farmer because I want to push against the boundaries that also exclude other people.

All farmers will benefit from such dismantling, and from the re-creation that follows. Feminism is my lifeline; it is a tool I use every day as a farmer. It was forged by all the people who think I can't do this. As I cultivate a vision of the feminist world we can create, my spirit strong and enduring, I say to the doubters, "Watch me."

FEMINIST FARMER

Nikiko asks me, "Are you a feminist farmer?" Answering the question makes me think of my history.

Growing up, I witnessed Japanese Americans never using their names on their fruit labels out of fear their produce would be rejected. We knew we didn't belong, but it seemed that the more we were marginalized, the harder we worked to belong.

And now Nikiko tells me she doesn't belong and may not want to "fit" into a narrow definition of the American farmer. She wants to create her own identity and meaning. As a queer woman. Of mixed race. As a worker and thinker.

Is this different than when I too returned to the farm? I left a possible future in academics, the lure of big cities, and a career in anything but farming. Yet I returned to the Japanese American community and then managed to open doors into the world of organic farming and specialized crops.

I had to renegotiate my identity. Once, Dad and I had a conversation about expanding the farm by buying land from a Japanese American down the street; we could work out a sweet deal. My uncle was also expanding his operation and suggested I do the same. Getting bigger was how you succeeded, how you made a name for yourself, how you established yourself. How you became a man. Marcy and I chose the opposite: to stay small, focus on doing what we do best, feed our passion for flavor and taste. Our farming does not belong in the world of business but rather in the world of food.

Now I see the connection between feminism and food: it means dismantling the meaning of what it is to be an American farmer. Nikiko is helping to change the narrative, helping us to reimagine food not as a commodity but as part of a movement to construct a world where justice and equality are part of the flavor and taste of our peaches.

Along the way, I learn more about our family's history as immigrants and non-Christians in an alien land. We have never belonged and perhaps never will. This is not such a bad thing: it feeds a spirit of creativity and innovation.

So call me a feminist farmer, one who works to carve a place in the land and claim a space in the long line of farmers. I seek to build a new network of relationships, to celebrate differences, and to forge new identities. I want to watch, witness, and learn from the next wave.

FARM DOGS

We have always had a farm dog. The first I can remember was appropriately named Dusty. Pat her short fur and a cloud of dust always puffed into the air. We did not know her breed (farmers often do not pay attention to pedigree). We just wanted a creature who would bark at strangers, be kind to kids, and like being outdoors. We also wanted watchdogs to guard us, and likewise we sought to become their guardians. I've always envied their free-spirited romps through the fields followed by naps in the warm sun or the cool shade—all while their owners work.

We were never very creative with names for our dogs. Over the decades, we have had about six different "Homers," no relation to the Greek writer, but a very common name—perhaps because they stayed close to home? We never used some of the most common names like Bailey, Bella, Max, or Lucy. We inherited names like Jake and Cody from rescue dogs or from dogs given to us by friends, or we went with ethnic names like Botchan (after a character in a famous Japanese novel) and Takara (the Japanese word for "treasure").

These farm dogs live outdoors, trot alongside pickup trucks, follow the farmer out to work. One of our golden retrievers, when he first arrived on the farm, loved to run and follow me out on the tractor. He didn't understand why I turned around at the end of one row to work the next, but, with blind loyalty, he ran alongside the tractor and faithfully jogged up and down the first twenty rows. Then, as fatigue settled in, he stopped accompanying me to the end of every row and plopped down in the middle, waiting for the tractor to return one row over. Finally, he became exhausted and limped home.

All of our dogs also hunt. Rabbits, quail, squirrels, the occasional pheasant, mice, and even lizards are fair game. Mostly, they just love chasing them. I enjoy watching them dash and dart, yelping with excitement. They don't stalk their prey like experts who were trained or selectively bred to perform. Instead they behave like good old dogs, unattached to humans, oblivious to the whole "I'm your master and you're supposed to serve me" role. Instead, when they get lucky with hunting and are gracious enough to share, they bring home the bounty and deposit it on the front doorstep.

And of course there's digging. Dirt flying, front paws burrowing, snouts stuck into gopher holes. Other times it's a slow, methodical scraping to carve out a cool spot to plant a warm body. Since farm dogs typically are not fenced, none of our dogs employ their excavating prowess to escape, but a number of young dogs have gotten into trouble tearing up home gardens, and some even felt compelled to exhume ancient bones. A few times, I'd discover (usually too late) that they had dredged new, creative channels for my irrigation water, the precious liquid running in places it wasn't supposed to. I like to think they were just trying to help.

Most farm dogs share a common trait: loyalty. They accompany their owners, following us into the fields, greet us with honest excitement each morning, and their love is unconditional. They are faithful and devoted.

Most of the time.

They always have the option to not join us in the fields and instead use common sense to stay in the cool shade on a 110-degree day. They are free to do as they please, as if adopting a Buddhist perspective of non-attachment and having no master. In fact, I am the one who fights to control nature by working in such heat. I recall driving out in my truck to check on a field and glancing in the rear view mirror to see if I had company. The dogs looked up, seemed to reflect, then put their heads back down and probably sighed before closing their eyes for another afternoon siesta.

I know stoic farmers, hard working and reserved, who usually hold their emotions in check. Yet they talk to their dogs with immense emotion and affection. I'm sure their other family members envy this attention.

Pets can display a type of absolute love and joy, and that's why they make such great companions. I love watching farm dogs play in the water. Each time they do it, they act as if it's their first moment of discovery. They explore, touch, jump, splash, leap, roll, bathe, cool, lie, rest. I want to join in and live the way they do, clear of limits, simple and free.

But then there are the times when dogs act like, well, dogs. They refuse to disclose the whole story when they come home with a slight gouge on their head after having disappeared for the night—probably out drinking and chasing coyotes. Sometimes they reappear with a special dog aroma, their scent betraying time spent in a neighbor's manure or compost pile. Instead of a confession, they stare at me with those puppy-dog eyes.

Sometimes dogs definitely act like dogs. They like to roll in feces. Why? One theory says it's instinctual behavior that helps hunters mask their own scents in order to sneak up on prey. Fortunately, the stink warns me to avoid petting or hugging them right after this bonding of theirs with their primal character.

Also, farm dogs will seek out and then "wear" the scent of the dead. Out in the countryside, they find things in varying stages of decay, rub their backs in it, and bring the odor home as if to announce where they've been. Then they seem to brag about their aromatic discovery and their new, stimulating perfume. Someone once claimed that the fatty oils they pick up from rotting things helps their fur's appearance, but I believe you have to be another dog to truly appreciate such a beauty regime. (I've seen other dogs deftly sniff each other during such exchanges.)

Our dogs like to eat our weeds and grass. They do have a discerning taste: I've never seen any of them eat Johnson grass, which may have a toxic flavor. They hate it and so do I—an alien species with few friends. But dogs do eat other types of grass and then vomit them up. It may be their natural way of taking care of an upset stomach, since they can't digest the plant fibers. It might provide them a quick fix—but it's certainly not pleasant to watch.

Farm dogs remain trusty and can become lifelong companions. And for old-time farmers like myself, who often work in solitude, they may be the only friend we interact with daily for hours and hours.

I wonder what dogs Nikiko will bring to the farm and eventually bury alongside older generations of farm animals. Will her dog choices reflect a different relationship with farming? Will she choose smaller and smarter dogs, ones good at herding and hunting? Or dogs that can spend more time inside—part of her effort to blur lines of division between farm and home? Which dog breed represents a classless world? A rescue dog? One that is of mixed race?

Or will she cross over into the world of cats? That would open a new universe—a world of peace, clear of divisions, of dogs and cats living together in harmony. That would take a talent and patience to train them that I don't have.

In my lifetime, I've buried my share of farm dogs. It's part of the responsibility of farm dog ownership—letting go of old friends but keeping them on the home place, eternal reminders of their presence. We reenact an age-old farming practice: we take care of the dead and perform a ritual of detachment, letting them go as we return their bodies to the earth we work.

As I grow older, I begin to think of my age in dog years. Our dogs will age more rapidly with each passing season, and so will I. I accept the fact we all will pass from this piece of land we call home. Together we will be forgotten. Dogs can teach me how to be free.

NIKIKO'S FIELD NOTE: TATTOOS

My mom hated tattoos. I think the pain of the process combined with the stereotypes about people who have tattoos always made it difficult for her to imagine the wide spectrum of reasons, thoughts, rituals, and identities intertwined with wanting to make a permanent mark on one's body. My experience with tattoos has been linked to spiritual understanding of life and those recurring existential questions I like to ponder.

When I got my first tattoo (a peach), it was to celebrate my decision to come back to the farm. It was my senior year of undergraduate work, and while fellow students were looking for post-graduation internships or jobs, or applying to graduate school, I was going home to farm.

After the bandage was removed, I remember looking down at my leg for the first time, at this beautiful peach now embedded in my skin, and thinking, "Well, hello. Why weren't you there before?" It felt less like an irreversible decision and more like a manifestation of part of me that had been growing under the surface all along.

One of my mother's complaints about tattoos is their permanence. Yes, it's true, my tattoos will be with me for the rest of my life, but that's part of their wisdom. My peach tattoo is a representation of something impermanent, just like its canvas. My body, as a canvas and vessel of my life, is mortal. With my death, my tattoo will also disappear. Just like the peaches we grow each year are consumed to feed us, I, one day, will disappear and my body will be returned to the cycle of growth and decomposition. It's freeing to carry reminders of this on my body. It helps to high-light both the importance of my existence and its small part in the larger universe of life and death.

Plus, it feels a little badass too.

LEANING

I pause from my shovel work and rest my hands on top of the wooden handle. I lean over and allow the trusty tool to brace me for a moment, giving my legs a rest, bringing my body into balance. I find comfort in the support. I am at peace.

My eyes pan the landscape. I can see the trees I planted with my father and the vines I planted with Marcy when we were first married. I see a block of older vines and the place Nikiko and I are planning for a new orchard as we plant her dreams while I can still work my shovel.

An old Japanese neighbor, typically very stoic, once shared a story when I was visiting his farm. I had studied him while he pruned his bonsai trees, watching his every subtle motion. Later, we stood silently in the shade of a tree near the back door. He asked me if I knew the Japanese kanji symbol for *hito*, which means "person."

He drew the character in the air with his finger. "You know, hito, written in two strokes. A long one with a short one holding it up." I followed the deliberate movement of his fingers and hand, a dramatic wave, a flick of the wrist, a dance in the air.

He repeated the performance, his hand in the air like a brush, motioning in a long, diagonal line dropping to the left, then a short one starting at the middle of the first stroke and angling to the right, sort of like an inverted Y, but with more graceful lines.

I could imagine the black ink of kanji, the tail of each stroke, and the snap of the old farmer's wrist as his brush lifts off the page, leaving an impression. I imagined his bold artwork on the side of his faded red barn, the black paint splashed on, leaving a spotted trail splattered across the lumber. I pictured a huge kanji character painted by this small, gentle old farmer, the drama made more powerful by the contrasts.

The old farmer repeated the brush strokes, a long line with a short one holding it up. "Hito," he concluded. "People. A farmer can't stand alone, has to lean against someone. Today, that's the way it is."

As the decades passed, I've grown more independent, not wanting to rely on others. The world of modern production agriculture has evolved, too: it's less personal and more about business. We no longer farm communally.

The irony is, of course, that as I age I must depend on others more and more. We have built our farm around a team—from the farmworkers to our distributors and end users (we call them "eaters"). I source materials to farm organically from a community of suppliers. Once relationships are founded, they last for decades.

After my father had his strokes, I witnessed him struggle as he changed from a self-reliant man who often worked alone to an aging man who needed assistance. He was still proud, but he was forced to accept help. Ironically, this freed him. By acknowledging his limits, he allowed us to help, without reservations, without feeling a need to reciprocate. He and we could show our love.

I worry about whether I'll be as accommodating as my father, whether I'll be able to recognize the changes that come, and to acknowledge that much of what I am now I may no longer be as I get old: will I affirm what life has to offer as I age?

As crude as it sounds, I recall helping my father take care of his bodily functions in his final years. We never spoke about it, but I knew he never was comfortable when I helped him go to the bathroom. I had removed the bathroom door in his house so his wheelchair could fit through the entryway. I'd park him and help shift his frail body to the toilet. I'd then stand in the hallway outside the light plastic curtain I'd installed over the doorframe as he grunted and groaned. He then signaled me to return. Despite all our best intentions, he still felt he had lost some of his dignity.

Will I be different? I joke uncomfortably with the kids: "Someday one of you might have to wipe my ass." That may be the ultimate act of sacrifice, when you learn to lean on someone. Is this yet another rite of passage shared between an old farmer and a new one? One more step of succession and ascension, one more part of letting go.

NIKIKO'S FIELD NOTE: EMPTY BINS

Dedicated to Uncle Alan

Each year, I'm mesmerized by the bins that hold the raisin crop. They are cavernous wooden raisin bins so big four people could fit inside—like huge sake boxes, each holding eight hundred pounds of raisins. All the bins are marked, many with family names painted on the sides. Diaz, Hiyama, Singh, Boorojian, Ludtke, Shapbazian, Oliver, Yamamoto, Six Jewels, Sangha. These are names from rural places often invisible in the urban food chain. These are the names of my neighbors, names of farmers, names of landowners, names of land leasers, names of immigrants, names of families tied to the land.

We reuse the bins year after year, and I wonder how many thousands of people have eaten raisins that have sat in them. How many workers have leaned their bodies into the sides, have pulled and pushed when the bins got stuck? How many splinters have lodged themselves in fingertips and then traveled to homes where loving caretakers removed them, kissed the wounded finger?

Our bins are older than I am. A few years ago, my dad repainted our name on the sides—"M A S U M O T O," in letters like the logo for the show *MASH*. The original print had faded, the black receding into the wood panels. For the new letters he chose purple. My favorite color.

One day after the purple re-paint, he got a call from the raisin plant that there were extra bins for us, left abandoned in a corner of the lot. Dad was confused; he keeps careful track of our bins and knew that we had already claimed all of ours for the season. I remember he left in the big truck to pick up the mystery bins, fairly certain the trip was a mistake and a waste of time.

I recognized the low rumble of his truck coming down the driveway, and when he rolled in, to all of our surprise, there were four bins strapped to the back. They did, indeed, have our name on the bins: "Masumoto," but not in purple.

They belonged, we learned, to my great uncle, who had farmed just a couple miles from our house. Uncle Alan used to go over to my jiichan and baachan's house for coffee and would sit with my jiichan and talk about the weather and hard times. He and my jiichan shared equipment and knowledge, advice and empathy, brother and brother.

When Uncle Alan passed away prematurely from a stroke years ago, the bins had been forgotten at the raisin plant. No one had reclaimed them, and they had sat empty on the lot until an observant forklift driver noticed them.

I walked out to the shed where my dad was painting purple "M A S U M O T O" onto the sides of Uncle Alan's bins. There were no other farmer heirs to pick them up, no one else who could say, "Yes, we are Masumotos. Yes, we still work the land." It was just us. And now at least for another generation, we'd use them again, fill their empty spaces with purple raisins, heavy and sweet. They belonged nowhere else, to no one else. We belonged to them.

It's getting harder to rationalize farming raisins when water is scarce and economic forces don't reward the work as much as it does for other crops. I see the benefits of change, but it's hard to imagine our farm looking different.

And if we stop growing raisins, what will happen to the bins? Will the names be repainted, or will they fade and be lost to time? Who will claim the bins left standing alone if their heirs are no longer growers but just eaters, far from the fields?

What will become of these family names: Diaz, Hiyama, Singh, Boorojian, Ludtke, Shapbazian, Oliver, Yamamoto, Six Jewels, Sangha, and Masumoto in purple?

THE STORY OF WATER

"When the well is dry, we know the worth of water."
—Benjamin Franklin

A drought plagues our farm, our valley, our state. Conflict fills the dusty air. There is an inherent tension to farming in the arid West, where rainfall has always been limited. We cultivate a desert, and only with the help of modern irrigation systems that bring water from the snowmelt of the Sierra, or pumps that pull water from beneath our lands, can we work this land. We farm a miracle, a mirage.

I worry about the grammar of water in the West. Should I say "my" or "our" water? It sounds like a debate on political correctness, but it's not: it's about keeping my damned trees and vines alive. I don't have the wealth or political power to call it "my" water, and already California's Sustainable Groundwater Management Act is legislating a regional, rather than individual, approach to the use of water pumped from the earth. We will all be forced to think in the plural.

Water may be the determining element for Nikiko's farming destiny. We can't grow things without water. Water will direct her journey on this land, decide which crops she grows, shape her own sense of the future. Water will define her identity. Water is the driver of all nature.

Drought amplifies this significance, and we can no longer take water for granted. While I fix a leak in our farm's cement irrigation line, I think of water in new ways. I cringe at the wasted water that the seeps out of the hairline crack, a problem I noticed years ago but only now feel compelled to fix. I try to grasp my relationship with water; it's not just for economic benefit but must also be seen as a limited resource, something sacred and with value in itself. For me, water is personal and meaningful. It's story that matters.

You hear talk about political capital and economic capital, but I'd like to introduce the concept of story capital. How water is framed in the narrative of farming, how the backstory of water is told, will leave behind a legacy that will affect all our lives, from politics to policy to personal strategies for survival. For example, let's look at this statement: "A single almond requires a gallon of water to grow." These words create an unshakeable metaphor about water and agriculture and, here, specifically, about almonds. Right or wrong, correct or overly simplified, the statement sticks. Our use of water brands us and casts us in certain lights (which may be unfavorable). Our water stories are being written for us whether we like it or not.

Our current story begins with an inherent conflict and tension: we can't create new water, all we have is now. The current water wars divide us into tribes locked in struggles we know well in California: urban versus rural, coastal versus inland, cities versus agriculture, and all of this versus environmental demands and considerations. We claim water is priceless, yet we act as if it's cheap, until reality strikes and we realize it's a scarce resource. We have created farming production models falsely based on a uniform weather pattern for the last fifty years. Climate change will upend existing methods, and I hope the smartest among us will transform the industry with innovation. I'm reminded of the term "creative destruction," the economic theory that says structures revolutionize by ceaselessly destroying the old and incessantly creating the new: the driving force of change in our parched valley is nature and weather.

The setting of our modern California water story takes place in a geography that's alien to much of the public, and even to policy-makers. Few pay attention to the watersheds that divide our state into natural regions. Much of our water supply is invisible, buried beneath us in aquifers, a setting few of us can picture. In this era of big data and information, a simple map may be the smartest tool to tell the water story in a compelling way. Imagine a relief map in three dimensions, with ridges and rises and deep gashes and basins that you can feel with your fingertips to comprehend the groundwater beneath your home and community. Might this knowledge change our perception of where we live and how we view the water we bank on for the future?

The main character in our story remains totally unpredictable: weather. Weather gives life and takes it away. We pray for storms, beg for rain, and swear at high-pressure systems. Fickle nature dominates our lives in the arid West.

At times it feels like we are trapped in the drama of a Kafka novel. We find ourselves thrown into bewildering situations, overcome with terror of the unknown, sometimes ripped with confusion and overwhelmed by systems and bureaucracies we cannot hope to control or change. We suffer nightmares of desperation and disorientation, with no clear course of action and no escape. At times, helplessness permeates life, and pain, anguish, and blame become the norm. We are confronted with unpredictable nature, with questions that have no right answers.

Stories help capture the complex context of life. My goal is to help find new voices and ensure they are represented. My hope is that as the drought continues, we can find the voice that allows articulation of a common bond: we are all connected by water. Water, more than anything else, will bond father and daughter, challenging the common phrase "blood is thicker than water." On our farm, water is as thick as blood.

As we look forward, what becomes the theme of our story of water? We are living in a water world of constant struggle amid a timeworn clash of haves and have-nots. The pain and fear are real; the anger and frustration are part of the new normal. Unfortunately, only when people turn on their faucets and find nothing there will they engage; then water becomes personal. Our plight will transform our communities. We must and will adapt, finding new compromises with new ethical and moral dimensions. Solutions will not be found solely by means of technology or politics: we will survive only with a whole new culture of water that impacts our daily lives.

Finally, what kind of story are we telling? Are we in a comedy or a political satire (how else can you cope with the politics of water)? Is this a drama? Maybe it's a tragedy in which the biggest decisions are made in times of crisis. Our choices may turn out to be wrong and costly, like Australia's desalination plants, built during their drought, which now sit idle. Or might this be a mystery as we strive to understand how to work with nature under conditions of global warming? Perhaps our future is less about solving problems and more about living these mysteries.

The story of water is not a simple linear narrative. We face a wonderfully complex future. We are witnessing a revolution, and we must cultivate our natural resiliency. If we are to survive, wisdom will unfold over generations. It makes for a great story, and we are living this story with every drink.

BETTER WITH AGE

"Youth is wasted on the young."
—George Bernard Shaw

After farming for decades, I find some things are better with age: fine wine, quality cheese, shade trees, classic cars. Do farmers belong in this category too? How about writers?

At more than sixty years old, I fear that I've already peaked, that it's all downhill from here. This feeling is compounded by the fact that in my youth I was not a brilliant person, not well read, not the model farm kid, and I often maneuvered my way into doing the least amount of work possible. I was not the best, and I knew it.

Yet secretly I hoped I had hidden talents yet to be discovered. It would take a lifetime for them to mature, so I braced for success later in life. As a young adult, I started to work harder, both in the fields and at my writing desk, believing in harvests to come.

According to research by economist David Galenson, there's still hope for me. He explored high achievers and found there is no evidence that great success is only the domain of the young. Older folks too have their place. Perhaps my big achievement is right around the corner.

Galenson studied famous poets, painters, writers, and filmmakers and discovered they fell into two groups, each of which took a different path to success. In one group were the bold young innovators who made dramatic leaps early in life. The other group consisted of older artists, those who worked hard on a gradual trajectory toward achievement. Members of both groups created outstanding work, they just had different timelines. There is hope for me and the farm: we're just still maturing.

I think of a twenty-year-old flame-throwing baseball pitcher whose 90 mph fastball is untouchable. I contrast him with a player late in his career who can no longer rely on brute strength and speed but has spent years practicing his technique, learning an assortment of pitches, relying on his savvy to trick the hitter at the plate.

Youthful innovators conceive brilliant ideas by thinking them through in their heads first, after which they act. The idea matters the most—the rest is mere execution. They need no practice, and move forward with certainty and an air of confidence, bold, brash, self-assured.

I've seen some farmers run their farms this way. They focus on growing an enterprise with rapid profits and large returns on investments. They are smart, quick, cocky, and often achieve much in a short time. Their success is measured in dollars and productivity.

Late bloomers experiment and live lives of trail and error. They figure things out by doing first, then learning from their failures. They practice and are willing to keep testing and exploring until they get it right. Humility often marks their work, or sometimes it's a tenacious (or stubborn) work ethic, a desire to overcome obstacles no matter what. These farmers can walk the fields and count the many experiments that failed. I know of old farmers who took years to understand their dirt—the pockets of sand, the lay of the land, the slope of an uneven field, the history of one small three-acre block of trees that struggles because cotton once grew in that soil and robbed it of the nutrients. It took years to replenish what was taken. And the old farmer had years to give to these thoughts, to cultivate his mind, just as a great crop is the culmination of years of careful cultivation, of building a fertile interdependence of natural elements working in harmony—the soil, the microbes, the bacteria, the fungi, all part of an organic whole that supports a peach tree and her fruit.

Prodigies hit their peaks early and then risk a gradual decline; late bloomers tend to have more steady rises and peak near the end. They have different types of passion, like sex verses love. One climaxes early; the other works to build long relationships. One type of farmer picks the fruit early, trying to lock in early-season prices, to beat the neighbors in a race to the market. The other type delays, harvesting when the fruit says so, trusting that those who wait will be rewarded.

I've envied young writers who have burst on the scene with their bold ideas, their shifts from conventions, often working with innovative, abstract ideas. My own style feels older, grounded in concrete images and simple language, certainly not part of any new wave of innovation. Of course, rejection letters and poor book sales eat away at one's confidence. At what point do you accept the fact that you're a midlist author, good enough to keep writing but never with broad success? Perhaps farming and writing are compatible in this way: they share a masochism for failure.

I accept being called a slow developer, bumbling and hesitant in my growth. My pace will be slow, frustrated by a rigorous journey of trial and error, tackling the same problem over and over. Success will be rare, yet I'll continue, motivated by the pursuit and a hope that I will discover what I'm great at. The journey is what's important, not the end. At least that's what I keep telling myself.

It's also important to remember that the late bloomers are not simply late starters: they are late because they were not very good until late in their careers. They needed to ferment. Only with age, time, maturity, and failure could they transform. Prodigies achieve success rather easily; but for late bloomers, it's all hard. However, rejection is part of the process for late bloomers; youthful stars burst with rejection. My files are full of rejection letters; my harvest records document volumes of disasters, some caused by wicked nature, others self-inflicted. Much too often, I have reassured myself in my field notes and writing journal: "Wait until next year…"

I wonder how many late bloomers never got a chance. They were considered old, past their prime; their talents were prematurely judged. By age thirty or forty they were told, "You don't have what it takes." They began to believe they weren't born with "it" and thus it was too late—it was time to concede and give up. Institutional support systems, such as work or academic settings, rarely reward the late bloomers.

But farming might. I've concluded that our best-tasting peaches are harvested from twenty-to-thirty-year-old trees. These are "old-growth" forests by most current farming standards, which typically see orchards removed after ten or fifteen years to accommodate a newer variety of peach, one that's redder and with a longer shelf life. But to me, our fifteen-year-old trees are only in their adolescence. The teenage fruit respond in monosyllabic answers and won't listen to me. They like to stay out all night, don't get enough sleep, and are very prolific but produce large crops of inconsistent quality. About what you'd expect. But after about the twentieth year in the ground, the trees begin to mature in a different way. Perhaps because the root systems are now well established, or maybe because I understand better how to care for these trees. They may not produce as much, but they produce better fruit. They seem to listen more and, in turn, I pay more attention to them. We talk and have grown-up conversations.

Late bloomers require patience and tenacity. They need to make themselves available for success, know their shortcomings, and work to overcome them. They tend not to be fatalistic: they don't necessarily accept their fates as failures, and instead spend their lifetimes painstakingly honing their craft. Then they find new energy. Journalist Malcolm Gladwell summarizes this well: late bloomers' stories are invariably love stories.

There are limits to this model of achievement, and not all people are late bloomers. This concept is not an excuse for being lazy or for procrastinating. Success at any level is 10 percent talent and 90 percent hard work. As author Daniel Pink writes, "This is a theory of creativity, not a Viagra for sagging baby boomer self-esteem." I have to fight the urge to seek out an instant peach Viagra for our aging orchards (or myself?).

At times, I wonder if my writing could use some "pop," maybe a nice scandal to liven things up. I believe some editors would love it if an evil bank or huge corporation drove us to bankruptcy. Once, when my father got a strange rash, I remember thinking that this was my opportunity for instant fame: I'd expose the pesticide that's killing my father, first via a magazine exposé, then in a book, and eventually by way of an award-winning documentary. When the doctor warned my father to stay inside and away from potential harm, I witnessed the horror on my father's face: you never tell a farmer to stay out of his fields. My priority shifted to getting him back on the land. The movie would have to wait.

Late bloomers require something outside themselves: a support system. Because of the long, slow process of experimentation, they need to be surrounded by believers and patrons—spouses, extended families, partners, good neighbors, lifelong friends, interested parties—who are willing to support the development of a slow-growing talent while the world rushes by in search of the young genius. These guardians back their late bloomers and are willing to see them through the journey. They offer trust and love. Our farm has provided me this as I've grown old. I wake exhilarated to begin another day of work. The farm gives me stories, and stories help me understand this life. The farm is one of my most treasured friends.

What does all this mean for Nikiko? Is she an early bloomer? She has already had great success in college and community organizing and working in the arts. She's a very fast learner, and yet she seems to understand the basic timeline of our farm: success is measured in decades, and now in generations.

So there's hope for us old dreamers. We still need the young, creative thinkers, and I want to support their brash insights. (Don't we all wish were young again at times?) But please leave a place for those of us who have worked hard for years and years. Our most innovative times may lie ahead. This is a world for both hares and tortoises. And for us tortoises, the best harvests are yet to come.

NIKIKO'S FIELD NOTE: HEARTFELT

I sat in a waiting room at the hospital feeling very out of place. There were two large families in the room, and new members kept arriving, greeting the others with long hugs. Many of them had puffy eyes, red from crying. I felt awkward. These families were clearly experiencing distress; people they loved were in danger. Nothing was certain except that the word "death" was unutterable.

I wondered if I should put my laptop and books away out of respect. After all, I wasn't worried. We were so nonchalant about my dad's planned procedure—an angiogram—that I had come to the hospital as the lone family representative, and I'd brought some work so I wouldn't lose a whole day waiting.

My dad was in one of the mysterious rooms behind the double swinging doors that led toward the operating facilities. I looked at my watch. He was probably shaking off the last of his unconsciousness by now. I'd take him home and he'd rest for a few days and, without skipping a beat, he'd be back to normal.

His doctor came out then and called for any family members of David Mas Masumoto. I stood and greeted him with a relaxed smile.

"We couldn't do the procedure because your dad has an artery that's completely blocked," the doctor reported.

I must have blinked twenty times. What? Whoa, that wasn't what I was expecting.

"He's going to need open-heart surgery."

Kaboom.

It was such a shock, I almost laughed.

Not so different now, I thought, looking at the families huddling together in the uncomfortable chairs. I wasn't sure how to make sense of it all in that moment, but I knew my dad was not going to be happy. This would mean he'd miss more work, and, I shuddered, wait, this would mean that I would have to do spring thinning by myself. Alone in the waiting room, and then alone in the fields without my teacher, my dad, my boss. What if something happened during the surgery?

I tried to calm myself and think positively. At least we caught this problem, at least this surgery would be preventative, not an emergency procedure. We had time to plan. We had time to strategize. He wasn't going to die.

A few weeks later, after the open-heart surgery, my dad was in the hospital for several days. It was weird. He couldn't do anything by himself at the beginning, and there were lots of tubes and electronic monitors beeping constantly, as if to say, "Still alive, still alive, still alive."

As he healed, he started to walk again. My brother and I happened to be visiting at the same time when a nurse came in to check on my dad and ask if he wanted to walk. Walking, my dad had read, was an important part of healing quickly, so he was determined to walk as much as he could.

The nurse turned to us and asked if we wanted to help him. I jumped up. Yes.

Since he was still attached to various tubes and machines, one of us had to push the mobile IV stand alongside him. When the nurse asked who wanted to do it, my brother nervously volunteered me. I know he was afraid of hurting my dad. I was afraid too.

We slowly exited his room, my dad using a walker, my brother and I on either side. He took slow, deliberate steps. His breath was short, each inhale sounding scratchy, as if something was constricting inside his chest. He lifted each leg and placed it gingerly forward. We could tell that every small movement caused him moderate pain. We walked in silence. I turned to look at my dad's face as he concentrated on the path ahead. He looked so tired, and really old.

I remembered back to the first time I comprehended that my mom and dad were whole people, that they weren't magical beings who needed nothing more from life than to fulfill my wishes. The biggest realization was that they made mistakes too, that they had weaknesses. I remember feeling then like I had graduated to a new stage in which I understood that we would age together, all making mistakes along the way; they were just a few phases ahead of me. It was a powerful moment, a true rite of passage.

And now, next to my dad, my brother on his other side, I realized something else. This was the other side of the humanity of parents: the reality that they are mortal. The walker made an eerie sound, its screws and lightweight metal crunching together to announce every forward movement. This is what the future would hold, I thought. Dad was only going to get older. Sickness, disease, injury, and other ailments—these challenges awaited us. And I knew we would again have to make this formation: my brother and I, the next generation, flanking our dad as he struggled. It would be our turn to support him.

I took a deep breath. We were lucky we caught this, we were lucky this was all preventative, we were lucky he was expected to have a full recovery. *I still have so much to learn from him,* I thought. *Did I start farming too late?*

THE FARMING HEART

I never thought that the dirt I work would rescue me. The methodical, daily grinding pace of farming provided the antidote for a bad heart. I should have had a heart attack years ago, but a farming heart saved me.

First, the major details: In the spring of 2014, my heart doctors discovered blockage in three major coronary arteries. I then had triple-bypass heart surgery. I continue to heal and feel great.

But, then, I had always felt well. I had none of the classic symptoms, which, I have since learned, are not always present when one is at risk for a heart attack. I had no shortness of breath, no chest pains, no upper-body aches. Earlier that year, my family doctor had given me a sixtieth birthday present: a routine stress test. They detected a slightly irregular heartbeat, but only under physical exertion at the end of a twenty-minute treadmill run. A second round of tests included an ultrasound, which proved again that my heart muscle was strong. But during a nuclear stress test, for which specialists inject a small amount of radioactive fluid in your blood and then monitor your blood flow as you work out, they detected a blockage. Next came the angiogram that revealed some scary truths: two of my arteries were completely blocked, and a third was about 60 percent blocked. One doctor said I should have already had a heart attack. Another said I must have already had one but not known it. The blockages were years old, the result of gradual buildup, possibly due to family genes more than anything else. What protected me was having a farmer's heart.

I want to be remembered for having a good heart. Certainly a caring heart for family and community, one that displayed empathy and endurance, much like the Japanese cultural ethos embodied by the terms giri (obligation) and gaman (perseverance). A good heart is one that has accepted an obligation to endure and to work hard, to find ways not only to survive but also, in my case, to be a good son, neighbor, and father. And a good partner to Nikiko on the farm.

I have a working-class heart. It necessarily embraces the visceral world of labor. I work with my hands in the fields and accept the bodily nature of what's required to farm. I have a new tool at my disposal: a strong heart that doesn't shy away from work. It's necessary and natural to have a vigorous heart in order to perform the everyday chores of farming. My heart has been tested, and I believe it has passed.

I say my heart saved me, but more precisely I was protected by the collateral arteries. If I understand correctly, my heart had developed a highly functioning and robust "natural bypass" network, a backup system. My doctors believe that my blockage had happened gradually—a very, very slow buildup of plaque, possibly stemming from genetics and heredity—which allowed the numerous smaller arteries to grow and compensate for the blood and oxygen loss of the blocked main arteries. Over years and decades, these collateral arteries grew and kept my heart pumping almost normally.

Ironically, the very things that I occasionally claimed were killing me—the long hours in the fields, the constant battle with weeds armed only with a shovel, the continuous daily walking to monitor trees and vines, the eternal chore to fix things that broke or were damaged—these draining activities saved my life. I have no romantic notions of working the earth in khaki shorts and a white Panama hat. My work is dirty, protracted, and habitual. I constantly exercise and strengthen my collateral arteries through hard physical labor.

I live and work slowly, as nature intended, over seasons and gradual changes and challenges. I toil and grind with a methodical rhythm, trying to avoid the tensions of keeping score. I do not record "personal bests." My pace is organic. Farming is the ultimate extreme of having an emphasis on living, not winning.

Diagnosis and bad news feel sudden and abrupt and leave us wanting immediate cures and fixes. But most health problems are slow to develop, such as weight gain, increased blood pressure or cholesterol, and even plaque buildup in your heart's arteries. You rarely wake up and discover suddenly that your body has failed you; more than likely, it happened slowly and gradually.

That's why I believe in slow farming.

As I recover from heart bypass surgery, I cherish this moment. For one of the few times, my choices in life have been rewarded. I feel vindicated for making the decision in my twenties to come back to farm. My working-class ethos has provided me with a life-giving health plan. My return on investment was a strong body. I have been generously compensated and I have recouped all expenses. I have amassed collateral wealth, a savings account I drew from for years. It's rare to be rewarded for who you are.

I inherited peasant genes from my family, generations of poor farmers here in America and in Japan. But I will no longer talk of ancestry as a deficit; I will instead speak of it as an asset. Manual labor shielded me; my shovel was part of a healthy benefits package. Working right to live right becomes a new prescription that will prolong my life. This is a legacy I can pass down, an invisible treasure that until recently was totally overlooked. The value of farming goes beyond making money or growing rich harvests. Who would have thought growing luscious peaches would preserve my heart? Many others who do physical work profit from similar collateral benefits, but such rewards are not acknowledged. Because we work with our hands and are considered blue collar, our labor remains undervalued.

For the first years of Nikiko's life, I was the primary caregiver. I not only watched her learn to walk but also played an active role. Now she witnesses me relearning how to walk our fields. I don't have a broken heart but instead a good one, made stronger by the work we do. I can feel the life-giving pace with each beat of my heart. Imagine: farming can save a life. And my hope is to pass down these farming genes.

ACKNOWLEDGMENTS

Thanks to the *Fresno Bee* and *Sacramento Bee,* where many of these stories and essays first appeared in my monthly column. Special thanks to my past and present editors at the *Bee*: Jim Boren, Bill McEwen, Dan Morain, Stuart Leavenworth, and Gary Reed.

Also thanks to *Saveur,* where a version of the essay "Change of Season" first appeared.

And thanks to the folks at Heyday, especially Malcolm Margolin and Gayle Wattawa, for their support and work in publishing this book. Stories continue to drive our passion and excellence.

—D.M.M.

To my ancestors, who gifted me with soil, fire, and dreams, thank you. To all the feminist teachers in my life, whose wisdom is constant and powerful, thank you. To all the doubters who have questioned my life decisions and the very essence of who I am, you have made me stronger and more resolved; thank you. To the artists who nudge, push, free, and inspire my creative heart, thank you. To my friends, the world is better with you in it and my life is so much richer for knowing you; thank you.

To my family, who is the source of infinite support, thank you. To my dad, whose drive to capture experiences in the written word is boundless, thank you. To my mom, the unending source of happiness and light, thank you. To my partner in joy, adventure, spontaneous dance parties, and laughter, Nichola, thank you.

—N.M.

IN APPRECIATION

The Masumotos and Heyday are appreciative of the following people for their financial contributions in support of this book and of Mas and Nikiko's accompanying California Storytelling Tour. We couldn't have done this alone: thank you so much!

The Sugimoto Family Foundation
George, Ruri, Lisa, and Nathan

Mike and Randy Jane Bayard
Julie and Craig McNamara
Bon Appétit Management Company
Pei-Ru Ko and Real Food Real Stories
Carolyn Jensen
Jesse Cool, Carlos Canada, and the peach picking,
        preserving, and cooking team at Flea Street
William and Carol Gong
Rich and Krissy Ishimaru
Bill Fujimoto
Jim Dodge
Pam Penner
Karen Shinto
Jackie Ryle
Jeannie Linder and Paul, Jenny, and Anne Rempel
Louise Kinoshita
Toby and Sally Rosenblatt
Gordon and Susan Hayashi

Sean Carr and Irene Lee
Carole Takaki
Jude and Woolf Kanter
Bob Moriguchi
Stephanie and Erica Pearl
Debbie Tom McCray
Mary Reed in honor of Florence and Wes Thieleke
Yoshiye Yamagiwa
Bob Benedetti
Angie Tagtow
Robert Boro
Organically Grown Company
Linda Tayian Hurst
John and Valarie Feaster
Saúl Jiménez-Sandoval
Rich Shintaku
Bruce Sowder
Shondell Spiegel
Regina Vukson
Anonymous (2)
Ben Arikawa
Domingo, Leticia, and Diego Cano
Irene Christensen
Paul, Patience, and Darrow
Mai Der Vang and Anthony Cody
Greg Donato
Elizabeth Donsky
Dave and Yvonne Duer
Elinor Fox
Anna Mae Gazo

Lee Herrick

Allison Hindman

Jane's Kids

Paula Johnson

Josephine and Yoichi Katayama

Mary Kimball

Cindy, Bob, and Kerry Klein, and Jes Therkelsen

Fran and Ted Loewen

David Masunaga

Cheryl Hisatomi McNabb

Jane Mermelstein

Janice Iki Motoshige

Ben and Doreen Nakayama

Camille Pannu

Cathy Mieko Pettit

Pam Pigg

KC Pomering, G-Free Foodie

Charlie Proctor

Murleen Ray

The Riddell Family

Sue Sakai-McClure

Carolyn and Gary Soto

Isaac Stein

Ellen Steloff

Tracy Stuntz and Jefferson Beavers

Valerie Vuicich

Don and Laura Weaver

The Larry Yamada Family

Geri and Elliott Yang-Johnson

ABOUT THE AUTHORS

David Mas Masumoto has been writing about farming and food for decades, capturing the serene moments of farm work like the "Lizard Dance" and documenting the family story of immigration, surviving World War II concentration camps, and forging a bond with the land and the Central Valley. He's served on National Council for the Arts (National Endowment for the Arts Board), has been an Irvine Foundation board member, and was former chair of California Council for the Humanities.

Nikiko is an emerging writer and storyteller, and calls herself an "agrarian artist." Her work has echoed the themes of cultural heritage, collective memory, and a search for justice. Her dreams have filled her imagination with poetry and stories about a world full of better farming and better food. You can see some of her ideas on stage in her TEDx talk from 2015 or at the BCFN Youth Manifesto symposium in Italy from 2015 or her one-woman show from 2011. Currently, she's a Creative Community Fellow with National Arts Strategies. She's also a proud board member of the Alliance for California Traditional Arts and Western States Arts Federation, and she serves on the College of Arts and Humanities Advisory Board at California State University, Fresno.

HEYDAY
into California

ABOUT HEYDAY

Heyday is an independent, nonprofit publisher and unique cultural institution. We promote widespread awareness and celebration of California's many cultures, landscapes, and boundary-breaking ideas. Through our well-crafted books, public events, and innovative outreach programs we are building a vibrant community of readers, writers, and thinkers.

THANK YOU

It takes the collective effort of many to create a thriving literary culture. We are thankful to all the thoughtful people we have the privilege to engage with. Cheers to our writers, artists, editors, storytellers, designers, printers, bookstores, critics, cultural organizations, readers, and book lovers everywhere!

We are especially grateful for the generous funding we've received for our publications and programs during the past year from foundations and hundreds of individual donors. Major supporters include:

Advocates for Indigenous California Language Survival; Anonymous (3); Judith and Phillip Auth; Carrie Avery and Jon Tigar; Judy Avery; Dr. Carol Baird and Alan Harper; Paul Bancroft III; Richard and Rickie Ann Baum; BayTree Fund; S. D. Bechtel, Jr. Foundation; Jean and Fred Berensmeier; Joan Berman and Philip Gerstner; Nancy Bertelsen; Barbara Boucke; Beatrice Bowles; Jamie and Philip Bowles; John Briscoe; David Brower Center; Lewis and Sheana Butler; Helen Cagampang; California Historical Society; California Rice Commission; California State Parks Foundation; California Wildlife Foundation/California Oaks; The Campbell Foundation; Joanne Campbell; Candelaria Fund; John and Nancy Cassidy Family Foundation; James and Margaret Chapin; Graham Chisholm; The Christensen Fund; Jon Christensen; Cynthia Clarke; Lawrence Crooks; Community Futures Collective; Lauren and Alan Dachs; Nik Dehejia; Topher Delaney; Chris Desser and Kirk Marckwald; Lokelani Devone and Annette Brand; J.K. Dineen; Frances Dinkelspiel and Gary Wayne; The Roy & Patricia Disney Family Foundation; Tim Disney; Doune Trust; The Durfee Foundation; Michael Eaton and Charity Kenyon; Endangered Habitats League; Marilee Enge and George Frost; Richard and Gretchen Evans; Megan Fletcher; Friends of the Roseville Public Library; Furthur Foundation; John Gage and Linda Schacht; Wallace Alexander Gerbode Foundation; Patrick Golden; Dr. Erica and Barry Goode; Wanda Lee

BOARD OF DIRECTORS

GETTING INVOLVED

To learn more about our publications, events and other ways you can participate, please visit www.heydaybooks.com.